INTERNATIONAL CENTRE FOR MECHANICAL SCIENCES

COURSES AND LECTURES - No. 293

KINETIC THEORY AND GAS DYNAMICS

EDITED BY

C. CERCIGNANI
POLITECNICO DI MILANO

SPRINGER-VERLAG WIEN GMBH

Le spese di stampa di questo volume sono in parte coperte da contributi
del Consiglio Nazionale delle Ricerche.

Originally published by Springer Verlag Wien-New York in 1988

ISBN 978-3-211-82090-2 ISBN 978-3-7091-2762-9 (eBook)
DOI 10.1007/978-3-7091-2762-9

PREFACE

The spectacular space programs developed in the late 1950's and in the 1960's produced a considerable interest in rarefied gas dynamics and led to the study of a great number of theoretical and experimental problems concerning flows of both neutral and ionized gases. Then the pace of the research activity slowed down considerably as a result of the de-emphasis of the space effort. Many important research groups who had contributed strongly to the rich harvest of results collected in the field of rarefied gas dynamics during the 1960's reconverted to somewhat more mundane areas, while others discovered important but less spectacular applications of the concepts and methods of kinetic theory in different areas.

Currently there is a renewed interest in high altitutde aerodynamics, reminiscent of the intense effort during the above mentioned period. Among the factors which contributed to the resurgence of this area one can mention the access to space provided by the U.S. Space Shuttle, the new applications in the 1990's and beyond foreseen by the U.S. Space Program, and the projected launch of the European space shuttle HERMES.

In the meantime mathematicians became interested in the theory of the Boltzmann equation and a significant part of a mathematical theory of this equation, which rules the time evolution of the distribution function in a gas according to kinetic theory, was developed.

Thus it was a timely decision that of having a CISM Course devoted to Kinetic Theory and Gas Dynamics *in June 1986. Seven lecturers, H. Babovsky, R.E. Caflisch, C. Cercignani, J.F. Clarke, W. Fiszdon, A. Palzewski and H. Spohn presented different aspects of kinetic theory and its applications to rarefied gas dynamics to a qualified audience of thirty "students", who ranged from recent Ph.D.'s to Full Professors.*

This volume contains the texts of six sets of lectures. Unfortunately Professor W. Fiszdon was unable to prepare his manuscript for this volume.

Hence the reader will find six chapters devoted to some mathematical aspects of the

Boltzmann equation (by C. Cercignani) and the Vlasov equation (by H. Babovsky and H. Neunzert), on existence and uniqueness theorems for the Boltzmann equation (by A. Palczewski), on the asymptotics of the Boltzmann equation and its relation to fluid dynamics (by R.E. Caflisch), on Physico-chemical Gas-Dynamics (by J.F. Clarke) and on the derivation of kinetic equations from Hamiltonian dynamics (by H. Spohn).

In publishing this volume in the by now well-established series of CISM Lecture Notes, we hope that many scientists who wish to get acquainted with the problems related to kinetic theory will find this work useful.

C. Cercignani

CONTENTS

THE BOLTZMANN EQUATION: SOME MATHEMATICAL ASPECTS

C. Cercignani
Politecnico di Milano, Milan, Italy

ABSTRACT

After recalling the basic facts concerning the Boltzmann equation, a few recent developments are surveyed. Among these the trend to equilibrium, the H-theorem for polyatomic molecules, exact solutions in the space homogenous case and existence of affine homoenergetic flows.

1. INTRODUCTION

The motion of a collection of N "hard spheres" moving in a finite box and colliding elastically with the wall and with each other can be pictured as the motion of a point moving in a subset Ω of R^{3N} obtained from the Cartesian N-th power of the box by deleting the subsets

$$|\underline{x}_i - \underline{x}_j| < \sigma \qquad i\neq j; i,j=1,\dots,N$$

where σ is the diameter of the spheres. The representative point moves in a rectilinear path and is specularly reflected at the boundary of Ω, see |1|,|2|. It turns out that the triple collisions are points of discontinuity of the flow, but these are negligible sets in a measure-theoretical sense, see |1|, |2|, |3|. If we want to describe the state of the system in probabilistic terms, we can introduce the N-particle distribution function

$$(1.1) \qquad F_N = F_N(\underline{x}_1,\dots,\underline{x}_N,\underline{\xi}_1,\dots,\underline{\xi}_N,t) \equiv F_N(P_N,t)$$

such that, for any subset of R^{6N},

$$(1.2) \qquad \int_E F_N d\mu_N = \text{Prob}(P_N \in E), \qquad d\mu_N = \prod_{k=1}^{N} d\underline{x}_k d\underline{\xi}_k .$$

One simply has

$$(1.3) \qquad F_N(P_N,t) = F_N(T_t P_N,0),$$

where T_t is the flow defined above. Accordingly, if $F_N(P_N,0)$ belongs to L^1, so does $F_N(P_N,t)$, due to the well-known fact that T_t is measure-preserving.

Alternatively one can introduce the one-particle distribution function $f(\underline{x},\underline{\xi},t)$ and say that f satisfies the Boltzmann equation, see |2|. The problem of reconciling the two descriptions is a long standing one and it is now understood that the reconciliation, if possible, should be obtained in the Grad limit, see |4|,|5|,|2|. In other words, Boltzmann's description should become valid in some sense when $N \to \infty$, $\sigma \to 0$ in such a way that $N\sigma^2$ tends to a finite limit.

A way of obtaining such a result has been indicated in a paper by the author |1| where the following result is obtained:

THE FACTORIZATION THEOREM. If

$$(1.4) \qquad F_N^s = \int F_N(P_N,t)d\mu_{N-s}, \qquad d\mu_{N-s} = \prod_{k=s}^{N} d\underline{x}_k d\underline{\xi}_k$$

has a sufficiently smooth limit F^s when $N \to \infty$, $\sigma \to 0$ with $N\sigma^2$ fixed, and

(1.5) $$F^s = \prod_{k=1}^{s} F^1(\underline{x}_k, \underline{\xi}_k, 0)$$

at t=0, then, at any time,

(1.6) $$F^s = \prod_{k=1}^{s} f(\underline{x}_k, \underline{\xi}_k, t),$$

where f is a solution of the Boltzmann equation corresponding to the initial datum $f(\underline{x},\underline{\xi},0) = F^1(\underline{x},\underline{\xi},0)$:

(1.7) $$\frac{\partial f}{\partial t} + \underline{\xi} \cdot \frac{\partial f}{\partial \underline{x}} = \frac{\sigma^2}{m} \int (f'f'_* - ff_*) | \underline{V} \cdot \underline{n} | \, d\underline{n} \, d\underline{\xi}_*$$

where $f_* = f(\underline{\xi}_*')$, $f_* = (\underline{\xi}_*')$, $f' = f(\underline{\xi}')$. This is the form of the Boltzmann equation for a gas of rigid spheres which will be used in what follows.

Here f is normalised in such a way that:

(1.8) $$\int f \, d\underline{x} \, d\underline{\xi} = M$$

Here m is the mass of a molecule and M the total mass. The meaning of f is an (expected) mass density in the phase space of a single particle, that is to say the (expected)"mass per unit volume" in the six-dimensional space described by $(\underline{x},\underline{\xi})$.

Please note that the unknown function f appears in the integral term not only with the arguments $\underline{\xi}$ (the current velocity variable) and $\underline{\xi}_*$ (the integration variable) but also with the arguments $\underline{\xi}'$ and $\underline{\xi}_*'$. The latter variables are related to $\underline{\xi}$ and $\underline{\xi}_*$ by the condition of being transformed into $\underline{\xi}$ and $\underline{\xi}_*$ by the effect of a collision,

(1.9) $$\underline{\xi}' = \underline{\xi} - \underline{\alpha}(\underline{\alpha} \cdot \underline{V}) \qquad (\underline{V} = \underline{\xi} - \underline{\xi}_*)$$
$$\underline{\xi}_*' = \underline{\xi} + \underline{\alpha}(\underline{\alpha} \cdot \underline{V})$$

Subsequently, O·Lanford |6| produced a proof of the existence of the limits provided $f(\underline{x},\underline{\xi},0)$ is smooth enough and one considers a sufficiently short time interval (0,T). The problem is justifying the formal deduction contained in |1| without introducing restrictions on the length of the time interval is still open.

In |1| an important role is played by the hierarchy of equation ss tisfied by the functions F^s. This infinite system, usually called the Boltzmann hierarchy, has been also considered by Lanford |6| and is discussed in some detail by H. Spohn |7| .

2. GENERALIZATIONS

Three possible generalizations suggest themselves: (1) molecules interacting with an at-distance force, (2) systems composed of several species of molecules such as a mixture of gases, (3) polyatomic gases.

The Boltzmann equations for mass points interacting with a central force can be written as follows |2-5|

$$\frac{\partial f}{\partial t} + \underline{\xi} \cdot \frac{\partial f}{\partial \underline{x}} = \iiint (f' f_*' - f f_*) B(\theta, V) d\theta d\varepsilon d\underline{\xi}_* \tag{2.1}$$

where $B(\theta,V)$ depends on the interaction law and is related to the differential cross section.

The next point to be discussed is the way of treating a mixture of different gases. The notation becomes complicated but there is no new idea, except for the obvious fact that we must derive n equations for the n one-particle distribution functions f_j $(j=1,\dots,n)$. The result is |5|

$$\frac{\partial f_j}{\partial t} + \underline{\xi} \cdot \frac{\partial f_j}{\partial \underline{x}} = \sum_{i=1}^{n} \frac{1}{m_i} \iiint (f_j' f_{i*}' - f_j f_{i*}) B_{ij}(\theta, V) d\theta d\varepsilon d\underline{\xi}_* \qquad (j=1,\dots,n) \tag{2.2}$$

where m_i is the mass of the molecules of the i-species, $B_{ij}(\theta,V)$ is computed from the law of interaction between the j-th and j-th species, and the arguments $\underline{\xi}'$, $\underline{\xi}_*'$ in f_j' and f_{i*}' in the i-th term are computed in terms of $\underline{\xi}$, $\underline{\xi}_*$, θ, ε from the laws of conservation of momentum and energy.

For the description of the behavior of polyatomic gases see section 4.

3. THE INITIAL AND BOUNDARY VALUE PROBLEM FOR THE BOLTZMANN EQUATION

The Boltzmann equation for monatomic gases:

$$\frac{\partial f}{\partial t} + \underline{\xi} \cdot \frac{\partial f}{\partial \underline{x}} + \underline{X} \cdot \frac{\partial f}{\partial \underline{\xi}} = Q(f,f). \tag{3.1}$$

where $f = f(\underline{x},\underline{\xi},t)$ is the distribution function, $\underline{x}$ and $\underline{\xi}$ are the position and velocity vectors of a molecule, $\underline{X}$ is the force per unit mass acting o, this molecule, $Q(f,f)$ the quadratic collision operator and t the time, is an integrodifferential equation containing partial derivatives and, as such, requires specification of initial and boundary data to be solved. Once initial and boundary conditions are given, an initial and boundary problem arises. This problem can be attacked from two viewpoints: one is to establish rigorous theorems of existence, uniqueness, continuous dependence on data, stability, etc. of the solutions of the problem; the other is to introduce techniques of obtaining exact or, more frequently, approximate solutions in a more or less explicit form. While it is only human that people working in one of these directions tend to praise the approach chosen and point out the scarce relevance of the other, it is fair to say that both approaches produce important information on the behavior of the solutions and, indeed, on the significance of the equation itself. This remark applies to all equations of mathematical physics, but it is particularly true in the case of the Boltzmann equation, whose significance and validity have been often misunderstood in the past. Particular solutions, even if approximate, serve the twofold purpose of describing in detail interesting experimental situations and strengthening one's faith in the adequacy of the mathematical model embodied in the Boltzmann equation. Rigorous theorems are important in helping to recognize whether the particular solutions are representative of more general case or whether they may be exceptional, as well as in determining whether or not there exist any solutions at all.

There is not much to comment about the initial data. Once the Boltzmann equation 2.1 is accepted as the evolution equation for $f = f(\underline{x}, \underline{\xi}, t)$ then one has to assign the values of f for $t = 0$. The assignment of these data is usually part of the physical description of the problem.

The situation is less obvious with respect to boundary conditions. This matter is frequently bypassed in general discussions by assuming a gas of infinite expanse or a bounded domain with periodic boundary conditions or, finally , a region bounded by solid bodies capable of specularly reflecting the molecules. These simplifications are ruled out when the flow past a solid body or within a region bounded by one or more solid bodies is studied. Then the boundary conditions describe the interaction of the gas molecules with the solid walls; it is to this interaction that one can trace the origin of the drag and lift exerted by the gas on the body and the heat transfer between the gas and the solid boundary. Unfortunately, both the theoretical and experimental information on gas-surface interactions are rather scanty; accordingly, one has to restrict oneself to general statements or to hypothetical models.

The basic theoretical concept needed for formulating the boundary conditions for the Boltzmann equation is that of the probability density $R(\underline{\xi}' \to \underline{\xi};\ \underline{x},\ t;\tau)$ that a molecule striking the surface with velocity between $\underline{\xi}'$ and $\underline{\xi}' + d\underline{\xi}'$ at the point $\underline{x}$ and time t re-emerges from practically the same point with velocity between $\underline{\xi}$ and $\underline{\xi} + d\underline{\xi}$ after a time interval τ (adsorption or sitting time). If R is known, it is easy to write the boundary condition for f |2,5|

$$(3.2)\qquad \begin{aligned} &|\underline{\xi}\cdot\underline{n}|f(\underline{x},\underline{\xi},t)\\ &= \int_0^\infty d\tau \int_{\underline{\xi}\cdot\underline{n}<0} R(\underline{\xi}'\to\underline{\xi};\underline{x},t,\tau)f(\underline{x},\underline{\xi}',t-\tau)|\underline{\xi}'\cdot\underline{n}|\,d\underline{\xi}'\\ &\qquad\qquad (\underline{\xi}\cdot\underline{n}>0,\underline{x}\in\partial\Omega), \end{aligned}$$

where $\underline{n}$ is the unit vector normal to the wall, which has been assumed to be at rest (otherwise, $\underline{\xi}$ and $\underline{\xi}'$ must be replaced by $\underline{\xi} - \underline{u}_0$, $\underline{\xi}' - \underline{u}_0$ throughout, $\underline{u}_0$ denoting the wall's velocity). $\partial\Omega$ denotes the boundary of the region Ω occupied by the gas.

Whenever the effective adsorption time $\tilde{\tau}$, number density n, range of gas surface interaction σ_0 and average normal velocity of the impinging molecules $\tilde{U}$ are such that $n\sigma_0^2\tilde{U}\tilde{\tau}\ll 1$ one can safely assume that $R(\xi' \to \underline{\xi};\underline{x},t;\ \tau)$ does not depend on the distribution function $f(\underline{x},\underline{\xi}'t)$; hence the scattering kernel R can be computed under the assumption that just one molecule of given velocity $\underline{\xi}'$ impinges upon the wall. If in addition $\tilde{\tau}$ is small compared with any characteristic time of interest in the evolution of f we can let $\tau = 0$ in the argument of f in the right hand side of eq. |3.2|; in this case the latter becomes

$$(3.3)\qquad \begin{aligned} &|\underline{\xi}\cdot\underline{n}|f(\underline{x},\underline{\xi},t) = \int_{\underline{\xi}'\cdot\underline{n}<0} R(\underline{\xi}\to\underline{\xi};\underline{x},t)f(\underline{x},\underline{\xi}',t)|\underline{\xi}\cdot\underline{n}|\,d\underline{\xi}'\\ &\qquad\qquad (\underline{\xi}\cdot\underline{n}>0,\ \underline{x}\in\partial\Omega) \end{aligned}$$

where

$$(3.4)\qquad R(\underline{\xi}'\to\underline{\xi};\underline{x},t) = \int_0^\infty R(\underline{\xi}'\to\underline{\xi};\underline{x},t;\ \tau)d\tau$$

Eq. (3.3) is, in particular, valid for steady state problem

If the wall restitutes all the gas molecules (i.e. it is nonporus and nonabsorbing), the total probability for an impinging molecule to be re-emitted, with no matter what velocity $\underline{\xi}$, is unity:

(3.5) $\int_{\underline{\xi}\cdot\underline{n}>0} R(\underline{\xi}'\to\underline{\xi};\ \underline{x},\ t)d\underline{\xi} = 1$

An obvious property of the kernel $R(\underline{\xi}'\to\underline{\xi};\ \underline{x},\ t)$ is that it cannot assume negative values:

(3.6) $R(\underline{\xi}'\to\underline{\xi};\ \underline{x},t)\geq 0.$

Another basic property of the kernel R, which can be called the "reciprocity law" or "detailed balance", is written as follows|2,5|

(3.7) $|\underline{\xi}'\cdot\underline{n}|f_w(\underline{\xi}')R(\underline{\xi}'\to\underline{\xi}) = |\underline{\xi}\cdot\underline{n}|R(-\underline{\xi}\to-\underline{\xi}')f_w(\underline{\xi}).$

where $f_w(\underline{\xi})$ is proportional to $\exp\left[-\xi^2/(2RT_w)\right]$, T_w being the temperature of the wall (in other words, $f_w(\underline{\xi})$ is a Maxwellian distribution for a gas at rest at the temperature of the wall). In eq. |3.7|, as well as in the following, the dipendence of R on $\underline{x}$ and t is not shown explicitly in order to make equations shorter.

We note a simple consequence of reciprocity: if the impinging distribution is the wall Maxwellian f_w and mass is conserved at the wall according to eq. |3.5|, then the distribution function of the emerging molecules is again f_w or, in other words, the wall Maxwellian satisfies the boundary conditions. In fact, if eq. |3.7| is integrated with respect to $\underline{\xi}'$ and eq. |3.5| used, the following equation is obtained:

(3.8) $\int_{\underline{\xi}'\cdot\underline{n}<0}|\underline{\xi}'\cdot\underline{n}|f_w(\underline{\xi}')R(\underline{\xi}'\to\underline{\xi})d\underline{\xi}' = |\underline{\xi}\cdot\underline{n}|\ f_w(\underline{\xi}) \qquad (\underline{\xi}\cdot\underline{n}>0).$

and this proves the statement. It is to be remarked that eq. |3.8|, although a consequence of eq. |3.7| (when eq. |3.5| holds) is less restrictive than eq. |3.7| and could conceivably by satisfied even if eq. |1.7| fails to hold.

As a consequence of the above properties, the following theorem holds|2,8,9|

Let C(g) be a strictly convex continuous function of its argument g. Then for any scattering kernel $R(\underline{\xi}'\to\underline{\xi})$ satisfying eqs. |3.5|, |3.6| and |1.8|, the following inequality holds:

(3.9) $\int f_w\underline{\xi}\cdot\underline{n}C(g)d\underline{\xi}\leq 0,$

where f_w is the wall Maxwellian, $g = f/f_w$ and integration extends over the full ranges of values of the components of $\underline{\xi}$, the values of f for $\underline{\xi}\cdot\underline{n}>0$ being related to those for $\underline{\xi}\cdot\underline{n}<0$ through eq. |3.3|. Equality in eq. |1.9| holds if and only if $f = f_w$ almost everywhere, unless $R(\underline{\xi}\to\underline{\xi})$ is proportional to a delta function.

The formal proof of the H-theorem rests on two inequalities: the first is due to Boltzmann and reads

(3.10) $$s = \int \log f Q(f,f) d\underline{\xi} \leq 0$$

equality holding if and only if f is (almost everywhere in velocity space) a Maxwellian; the second one is much more recent (|2,8-1|; see discussion in Refs. |2 and 9|) and has the following form (Corollary of Eq. (3.9)):

(3.11) $$\int f \log f \underline{\xi}\cdot\underline{n} d\underline{\xi} \leq -\frac{(\underline{q}\cdot\underline{n})\text{solid}}{RT_w}$$

where T_w is the wall temperature, R the gas constant, $\underline{q}$ the heat flow vector, $\underline{n}$ the normal unit vector pointing into the gas. The subscript "solid" means that $\underline{q}\cdot\underline{n}$, which is, in general, discontinuous at the surface, has to be evaluated on the solid side. Equation (3.11) holds if the values of f for $\underline{\xi}\cdot\underline{n}>0$ are related to those for $\underline{\xi}\cdot\underline{n}<0$ through Eq.(3.3). Equality applies if and only if $f = f_w$ (the "only if" part does not apply if $R(\underline{\xi}'\to\underline{\xi};\underline{x},t)$ is a delta function).

Some comments are in order here concerning the classical inequality (3.10) which is a trivial consequence of the properties of the logarithm and of the identy |2,5|

(3.12) $$\int \phi Q(f,f) d\underline{\xi} = -\frac{1}{4}\int (\phi'+\phi'_*-\phi-\phi_*)(f'f'_*-ff_*)B(\theta,|\underline{V}|)d\varepsilon d\theta d\underline{\xi} d\underline{\xi}_*$$

In order to prove this identity, one uses the fact that the absolute value of the Jacobian of $\underline{\xi}'$ and $\underline{\xi}'_*$ with respect to $\underline{\xi}$ and $\underline{\xi}_*$ for fixed $\underline{\alpha}$ is unity; a trivial consequence of the linearity of Esq. (2.3) and of the fact that the transformation expressed by these equations is its own inverse |20|. In addition, one exploits the fact that is θ' is the angle between $\underline{V} = \underline{\xi}' - \underline{\xi}'_*$ and $\underline{\alpha}$, then $\theta = \theta'$ as well as the trivial observation that changing $\underline{\alpha}$ into $-\underline{\alpha}$ does not matter. In order to conclude that

(3.13) $$\left|\frac{\partial(\underline{\xi}',\underline{\xi}'_w,\theta',\varepsilon')}{(\underline{\xi},\underline{\xi}_*,\theta,\varepsilon)}\right| = 1$$

it only suffices to remark that we are not obliged to use θ and ε as angles to identify $\underline{\alpha}$ (though we may find it useful) and hence we can keep $\underline{\alpha}$ fixed when we change variables from $\underline{\xi}$, $\underline{\xi}_*$ to $\underline{\xi}'$, $\underline{\xi}'_*$. To be more explicit:

(3.14) $$\sin\theta d\theta d\varepsilon d\underline{\xi} d\underline{\xi}_* = \sin\eta d\eta d\phi d\underline{\xi} d\underline{\xi}_* = \sin\eta d\eta d\phi d\underline{\xi}' d\underline{\xi}'_* = \sin\theta' d\theta' d\varepsilon' d\underline{\xi}' d\underline{\xi}'_*$$

where η and φ are polar angle indipendent of $\underline{V}$ and use has been made of the well-known rotation invariance of the area element of the unit sphe re. Equation (3.13) is a consequence of Eq. (3.14) and of $\theta' = \theta$.

We have spelled out the details of this argument because Eq.(3.13) is said to be the consequence "of direct if lengthy calculations" in the book by TRUESDELL and MUNCASTER |9|. The above proof seems to have been misunderstood if one can judge from a few letters of criticism from the readers of a book |5| where the proof itself is sketched. An indirect criticism without a precise reference seems to be contained in a paper by SCHNUTE |12| who remarks that it is not uncommon to infer that Eq. (3.13) is a consequence of the fact that the transformation $(\underline{\xi},\underline{\xi}_*,\theta,\varepsilon)\to(\underline{\xi}',\underline{\xi}'_*,\theta',\varepsilon')$ is its own inverse. This argument would indeed be false when applied to this (nonlinear) transformation, but it is correct when applied to the linear transformation $(\underline{\xi},\underline{\xi}_*)\to(\underline{\xi}',\underline{\xi}'_*)$ for a fixed $\underline{\alpha}$. Schnute's criticism has been echoed by TRUESDELL and MUNCASTER |9|.

Concerning inequality (3.11), we remark that if the scattering kernel is a delta function, i.e.

$$R(\underline{\xi}'\to\underline{\xi};\underline{x},t) = \delta(\underline{\xi}-\underline{\xi}(\underline{\xi}')) \tag{3.15}$$

where $\underline{\xi}(\underline{\xi}')$ is a (differentiable) function which we assume to be uniquely invertible, then Eq. (3.5) implies (adopting a reference frame where the wall is locally at rest):

$$\left|\frac{\partial\underline{\xi}}{\partial\underline{\xi}'}\right| |\underline{\xi}(\underline{\xi}')\cdot\underline{n}| = |\underline{\xi}'\cdot\underline{n}| \tag{3.16}$$

$$|\underline{\xi}(\underline{\xi}')| = |\underline{\xi}'| \tag{3.17}$$

The restriction of Eqs. (3.16) and (3.17) to $\underline{\xi}'\cdot\underline{n}<0$ can be dropped by defining $\underline{\xi}(\underline{\xi}')$ to coincide with its own inverse when $\underline{\xi}'\cdot\underline{n}>0$. All the transformations defined in the entire R^3 and satisfying Eq. (3.17) are known to be linear and to have a Jacobian with unit absolute value. If we restrict ourselves to the subset of these transformations for which Eq. (3.16) is also satisfied (the Jacobian factor may now be suppres sed), we find that

$$\underline{\xi}(\underline{\xi}')\cdot\underline{n} = -\underline{\xi}'\cdot\underline{n} \tag{3.18}$$

i.e. the linear transformation $\underline{\xi}'\to\underline{\xi}(\underline{\xi}')$ transforms $\underline{\xi}'\cdot\underline{n}$ into $-\underline{\xi}'\cdot\underline{n}$. It is otherwise a rotation in a plane tangential to the wall. If we add the somewhat natural restriction that $\underline{\xi}(\underline{\xi}')$ lies in the plane of $\underline{n}$ and $\underline{\xi}'$, we are left only with specular reflection and the parity transformation $\underline{\xi}' = -\underline{\xi}$, a result first discovered by SCHNUTE |14|. In either case Eq. (3.11) holds trivially because both sides vanish.

Once Eqs. (3.10) and (3.11) are established we have only to assume that f is sufficiently well behaved to conclude that in a bounded domain

$$H \leq \int_{\partial\Omega} \frac{(\underline{q}\cdot\underline{n})_{solid}}{RT_w}\, dS \tag{3.19}$$

where dS is the surface element on $\partial\Omega$ and

$$H = \iint f \log f \, d\underline{\xi} \, d\underline{x} \tag{3.20}$$

is the celebrated H-function (actually a functional of f) introduced by BOLTZMANN |13|. Equation (3.19) is an extension of Boltzmann's theorem and appears to have been proved formally in full detail for the first time in Ref. |8| (see, however, also Ref. |11| and |10| and discussion in Refs. |2| and |9|). Originally, the theorem was established with no boundary term at all or in the presence of specularly reflecting boundaries |14|. In unbounded regions, the condition $|\underline{x}|^2 f \log f |\underline{\xi}\cdot\underline{n}| d\underline{\xi} \to 0$ for $|\underline{x}| \to \infty$ has to be added for Eq. (3.19) to hold.

In order to make the proof of Eq. (3.19) rigorous, one might sufficient conditions for the formal steps to be correct. This would be easy to do but there is little advantage in it because the theorem might be true under more general conditions (perhaps in weak sense). Before discussing this matter further (sect. 5), we consider the case of a polyatomic gas.

4. ON THE H-THEOREM FOR POLYATOMIC GASES

Since the publication, in 1872, of the famous memoir of Boltzmann, |13| containing, as one of its many remarkable features, the proof of the H-theorem for monatomic gases, the debate on the meaning of this result has gone on and we can say that even today one can find scepticism about the meaning of this basic contribution of Boltzmann's nonequilibrium statistical mechanics |9|.

It is even more surprising, therefore, to find that the extension of the H-theorem to a gas whose molecules are arbitrarily complex structures, which was proposed by Boltzmann in the same basic paper |3| and subsequently criticized by Lorentz, |15| is not usually discussed in treatises on kinetic theory, even those which deal with explicit models for polyatomic molecules, |16| with the notable exceptions of Boltzmann's "Lectures" |17| and Tolman's book |18|.

As is well known, Lorentz's objection to Boltzmann's proof is based on the fact that one cannot expect to find any inverse collision for an arbitrarily chosen collision, with the exception of molecules having a spherically symmetric interaction. We underline the circumstance that it is not the geometric shape of the molecules which matters but rather the details of the interaction, as shown by the case of rough spherical molecules |19,20,16|.

Boltzmann tried to fight with the objections of Lorentz. He carefully distinguished between initial and final, time reversed and corresponding "constellations" finally arriving at the so-called cycle proof of the H-theorem | 17, 18|. It is not clear, however, that this argument really proves something; as recently as 1972, G.E. Uhlenbeck |22| stated that he had some doubts about these generalizations. He then suggested a way of escape in the quantum treatment of the collisions, as developed by L. Waldmann |29, 27| and R.F. Snider |25|. As a matter of fact the possibility of finding a proof of the H-theorem for polyatomic gases in quantum mechanics as a consequence of the unitariety of the scattering matrix had been pointed out as early as 1952 in a paper by Stueckelberg |26| and described later in a connection with a quantum Boltzmann equation by Waldmann | 27|.

It is the purpose of this part to point out that there is a proof of the H-theorem for a gas of purely classical molecules with arbitrarily complex structure. The starting point will be the reciprocity relation for the scattering probability; the latter holds because of the time-reversible character of the equations of the dynamics.

In this part we follow a paper by M. Lampis and the author |28|.

We shall deal with molecules described by a set of variables, which will include the position vector $\underline{x}$ and other variables which will be collectively denoted by $\underline{p}$ ($\underline{p}$ will be a vector in an n-dimensional space); the latter will include, e.g., velocity and angular velocity or angular momentum.

The distribution function $f = f(\underline{x},\underline{p},t)$ will satisfy an evolution equation of the form

$$(4.1)\quad \frac{\partial f}{\partial t} + \underline{P}\cdot\frac{\partial f}{\partial \underline{P}} = \iiint \big| f(\underline{x},\underline{p}',t)f(\underline{x},\underline{p}'_*,t)W(\underline{p}',\underline{p}'_*\to\underline{p},\underline{p}_*) - f(\underline{x},\underline{p},t)f(\underline{x},\underline{p}_*,t)W(\underline{p},\underline{p}_*\to\underline{p}',\underline{p}'_*)\big| d\underline{p}_* d\underline{p}' d\underline{p}'_*$$

where $\underline{P}$ is a n-dimensional vector which is assigned when the causes different from collision deviating the molecules from rectilinear paths are given through the equation of collision-free motion:

$$(4.2)\quad \dot{\underline{p}} = \underline{P}(\underline{x},\underline{p})$$

In Eq. (4.1) $\underline{p}_*$ denotes the variables of the partner of $\underline{p}$ in a collision, $\underline{p}'$, $\underline{p}'_*$ the variables of two molecules which end up in a state $\underline{p}$ and $\underline{p}_*$, respectively at the end of a collision. The transition "probability" $W(\underline{p}', \underline{p}'_*\to\underline{p},\underline{p}_*)$ is essentially the differential cross section multiplied by relative speed, and in fact one has

$$(4.3)\quad \iint W(\underline{p}',\underline{p}'_*,\underline{p}_*)d\underline{p}'d\underline{p}'_* = |\underline{\xi} - \underline{\xi}_*|\sigma_t$$

where σ_t is the total cross section. The latter could, in principle, depend on the variables $\underline{p}$ and $\underline{p}_*$; as a matter of fact we shall assume that it does not, by a sort of trick. First, we shall assume that no interaction takes place when the molecules are separated by more than a certain distance r_0 (finite range interaction). We can assume that σ_t is exactly πr_0^2 provided we allow the presence of no-scattering events in W: in other words if for certain values of $\underline{p}'$, $\underline{p}'_*$ there is no interaction between the molecules even at a distance smaller than r_0, we shall consider W different from zero there and actually proportional to a delta function ensuring that no change takes place during the "collision". This convention will notably simplify our treatment. The case of infinite range interaction can be dealt with, if desired, by a limiting procedure on the results which will be established here.

We remark that the time reversibility of the equation of motion implies that there is a transformation $\underline{p}\to\underline{p}^-$ (typically $\underline{p}^- = -\underline{p}$) such that

$$(4.4)\quad W(\underline{p}',\underline{p}'_*\to\underline{p},\underline{p}_*) = W(\underline{p}^-,\underline{p}_*^-\to\underline{p}'^-,\underline{p}'^-_*)$$

This property will be called as usual "reciprocity". In addition the transformation from $\underline{p}$ to $\underline{p}$ is measure preserving. In the case of a perfectly spherical interaction, the additional property

(4.5) $$W(\underline{p}',\underline{p}'_* \to \underline{p},\underline{p}_*) = W(\underline{p},\underline{p}_* \to \underline{p}',\underline{p}'_*)$$

called detailed balance, applies.

Use of Eq. (4.5) would allow a proof of the H-theorem by a word-by-word repetition of the proof holding for the monoatomic gas. If Eq. (4.5) fails, then one usually invokes Boltzmann's argument based on the assumption of "closed cycles of collisions", |21,17,18| which many authors have found difficult to follow. The main feature of the proof, however-viz., that many collisions have to be lumped together- is key to the proof in the quantum case as well as to the proof for the classical equation which we are presenting here. We integrate both sides of Eq. (4.4) with respect to $\underline{p}$, $\underline{p}_*$ to find

(4.6) $$\begin{aligned}\iint W(\underline{p}',\underline{p}'_* \to \underline{p},\underline{p}_*)d\underline{p}d\underline{p}_* &= \iint W(\underline{p}^-,\underline{p}_*^- \to \underline{p}'^-,\underline{p}'^-_*)d\underline{p}d\underline{p}_* \\ &= \iint W(\underline{p},\underline{p}_* \to \underline{p}'^-,\underline{p}'^-_*)d\underline{p}d\underline{p}_*\end{aligned}$$

where we have changed the integration variables from $\underline{p},\underline{p}_*$ to $\underline{p}^-,\underline{p}_*^-$ and then supressed the minus superscript, which is no longer needed ($d\underline{p}d\underline{p}_* = d\underline{p}^-d\underline{p}_*^-$). The last integral in Eq. (2.6) is nothing else, according to Eq. (4.3), than

(4.7) $$|\underline{\xi}^- - \underline{\xi}_*^-|\sigma_t = |\underline{\xi} - \underline{\xi}_*|\sigma_t = \iint W(\underline{p},\underline{p}_* \to \underline{p}',\underline{p}'_*)d\underline{p}d\underline{p}_*$$

where $\underline{\xi}^- = -\underline{\xi}$ and $\underline{\xi}_*^- = -\underline{\xi}_*$ and the constancy of σ_t have been used.

Equations (2.6) and (2.7) together give

(4.8) $$\iint W(\underline{p}',\underline{p}'_* \to \underline{p},\underline{p}_*)d\underline{p}d\underline{p}_* = \iint W(\underline{p},\underline{p}_{,*} \to \underline{p}',\underline{p}'_*)d\underline{p}d\underline{p}_*$$

This is a basic new relation which will be used in the next section to prove the H-theorem. The importance of a relation such as Eq. (4.8) was stressed by Waldmann, who, guided by analogous relation in the quantum case, wrote it for the first time in the particular case of a gas of classically rotating linear molecules | 29 |. He did not present, however, any proof of Eq. (4.8) but stated that "one must get the (purely mechanical) normalization property" expressed by Eq. (4.8) As a possible proof he seems to hint at a complete calculation with the simplifying assumption of "averaging over all possible phase angles before and after collision ". This averaging, though practically convenient in order to avoid the use of ignorable coordinates (in the absence of external electric fields), is not required in our direct proof. According to our approach, Eq. (4.8) is a general property following from the time reversibility of a microscopic equations of motion. It was proved for the first time, to the best of our knowledge, in Ref. |28|.

It is now a simple matter to prove Boltzmann's lemma, according to which we let

(4.9) $Q(f,f) = \int\int \left[f'f'_*W(\underline{p}',\underline{p}'_* \to \underline{p},\underline{p}_*) - ff_*W(\underline{p},\underline{p}_* \to \underline{p}',\underline{p}'_*) \right] d\underline{p}_* d\underline{p}' d\underline{p}'_*$

then

(4.10) $\int \log f\ Q(f,f) d\underline{p} \leq 0$

f', f'_*, f_*, in Eq. (3.1) denote, as usual, the distribution function f having, as p-argument, $\underline{p}'$, $\underline{p}'_*$, $\underline{p}_*$, respectively. This lemma is the basic prerequisite in the traditional way |13,16,17,2,5|.

In fact we multiply Eq. (4.9) by log f and integrate, we otain

(4.11) $\int \log f\ Q(f,f) d\underline{p}$

$$= \tfrac{1}{2} \int\int\int\int ff_* \log \left(\frac{f'f'_*}{ff_*} \right)\ W(\underline{p},\underline{p}_* \to \underline{p}',\underline{p}'_*) d\underline{p} d\underline{p}_* d\underline{p}' d\underline{p}'_*$$

This relation can be obtained by standard manipulations, i.e., suitable changes of variables and labels and does not involve any use of the properties of $W(\underline{p},\underline{p}_*\ \underline{p}',\underline{p}'_*)$. Equation (4.10) does not permit to use the standard argument for proving the H-theorem; to use the latter one should add to Eq. (4.11) the same equation after having interchanged primed and unprimed variables and use Eq. (4.5), which, however, does not hold in general.

At this point we use a mathematical trick which seems to have been suggested by Pauli |26| in connection with the quantum mechanical proof. Along with the identity expressed by Eq. (4.11) we consider the other identity

(4.12) $\int f'f'_*W(\underline{p}',\underline{p}'_* \to \underline{p},\underline{p}_*) d\underline{p} d\underline{p}_* d\underline{p}' d\underline{p}'_* = \int ff_*W(\underline{p},\underline{p}_* \to \underline{p}',\underline{p}'_*) d\underline{p} d\underline{p}_* d\underline{p}' d\underline{p}'_*$

which follows from a change of labels and expresses conservation of mass. In the left-hand side of Eq. (4.12) we can transform the integral over the unprimed variables according to Eq. (4.8) to obtain

(4.13) $\int f'f'_*W(\underline{p},\underline{p}_* \to \underline{p}',\underline{p}'_*) d\underline{p} d\underline{p}_* d\underline{p}' d\underline{p}'_* = \int ff_*W(\underline{p},\underline{p}_* \to \underline{p}',\underline{p}'_*) d\underline{p} d\underline{p}_* d\underline{p}' d\underline{p}'_*$

or equivalently

(4.14) $\tfrac{1}{2} \int ff_* \left(\frac{f'f'_*}{ff_*} - 1 \right) W(\underline{p},\underline{p}_* \to \underline{p}',\underline{p}'_*) d\underline{p} d\underline{p}_* d\underline{p}' d\underline{p}'_* = 0$

Hence we can substract the integral appearing in the left-hand side of Eq. (4.11) without changing anything. We obtain

(4.15) $$\int \log f\, Q(f,f)d\underline{p} = \frac{1}{2}\iiiint ff_* \left[\log\frac{f'f'_*}{ff_*} - \left(\frac{f'f'_*}{ff_*} - 1\right)\right] \times W(\underline{p},\underline{p}_* \to \underline{p}',\underline{p}'_*)d\underline{p}d\underline{p}_*d\underline{p}'d\underline{p}'_*$$

Since f, f_*, W are positive and

(4.16) $$\log x - (x - 1) \leq 0$$

the equality sign holding if and only if $x = 1$, we conclude the inequality (4.10) is proved, the quality sign applying if and only if (almost everywhere):

(4.17) $$f'f'_* = ff_*$$

This relation, as is well known, leads to the equilibrium distribution (log f must be a linear combination of the collision invariants).

5. RIGOROUS PROOFS OF THE H- THEOREM

We now examine the particular cases when the H-theorem has been rigorously proved. The starting point is, generally speaking, the proof of an existence and uniqueness theorem under suitable restriction on initial and, if necessary, boundary data. Then under the same or more restrictive conditions, the existence of H for all time follows, together with Eq. (3.19).

In the space homogeneous case ($\partial f/\partial \underline{x} = 0$), the right hand side in Eq. (3.19) as well as the integration with respect to $\underline{x}$ in Eq. (2.16) have to be omitted , of course. This is the only case in which satisfactory results are available.

It is true that the problem of a trend toward equilibrium of a spatially homogeneous gas is trivial from the fluid dynamic point of view because there is no change at all in density, velocity, temperature and hence no fluid-dynamics in the usual sense of the word, but interesting facts do occur in velocity space and the mathematics is by no means trivial.

The first existence and uniqueness theorem for this problem was obtained by CARLEMAN in the case of hard sphere molecules. He obtained a strong non linear result in the Banach space with the norm

$$||f||_{\alpha} = \text{Ess}\sup_{\underline{\xi}} \left| |f(\underline{\xi})| \, (1 + |\underline{\xi}|^2)^{\alpha} \right| \tag{5.1}$$

where $\alpha > 3$. Uniqueness and uniqueness in the large are shown in the cone of positive functions together with the existence of the H-functional as a non-increasing function of t and the tendency to a Maxwellian distribution for $t \to \infty$.

WILD |30| and MORGENSTERN |31| proved similar results for the simpler case of Maxwell's potential with angular cutoff, but their results are less complete than Carleman's as far as tendency toward equilibrium is concerned. The best available results for the space homogeneous case are those of ARKERYD |32| who was able to prove that, for a gas of rigid spheres or with angular cutoff, there is global existence of a positive solution if $f_0 \geq 0$, $(1+|\underline{\xi}|^2)^{\alpha} f_0$ and $f_0 \log f_0$ belong to L^1.

If α can be taken not to be less than 2, then the solution is shown to be unique and the assumption on $f_0 \log f_0$ is not needed for existence. If the latter hypothesis is satisfied, however, it follows that f log f is L^1 for any $t > 0$ and the H-theorem is valid; in addition if $\alpha > 2$, there is a trend equilibrium (i.e. f tends to a Maxwellian for $t \to \infty$) in the sense of weak convergence in L^1.

6. CONSEQUENCES OF THE H-THEOREM AND COUNTEREXAMPLES

If one accepts the H-theorem, he can draw some consequences. These consequences amount to saying that, under certain conditions, the distribution function will tend to a Maxwellian. The difficult point is to make these conditions explicit.

In the space homogeneous case and in the absence of body forces, the argument goes back to BOLTZMANN himself |14|. TRUESDELL and MUNCASTER |9| criticize the traditional "proof", especially in the form given by CHAPMAN and COWLING |16|. The argument under discussion starts from the fact that $\dot{H} \leq 0$ implies that H monotonously decreases in time unless f is a Maxwellian. It is then noted that H is bounded from below and H tends to a limit H_∞. This limit is asserted to correspond to a state of the gas in which $\dot{H} = 0$; this in turn is known to imply that f is a Maxwellian.

There are three difficult points in the argument: the boundedness of H from below, the fact that $\dot{H}$ tends to some limit (which must be zero) and the existence of a limit of f when $t \to \infty$.

In the space homogeneous case the first difficulty is easily disposed of. This aspect is discussed in some detailed by Chapman and Cowling who remark that, roughly speaking, the existence of the integral $\int f\underline{\xi}^2 d\underline{\xi}$ implies that of H. In order to make their argument rigorous, let us consider the elementary inequality

$$\text{(6.1)} \qquad f\log f \geq f\log f_m + f - f_m$$

where f_m is a Maxwellian having the same density, bulk velocity and temperature as f. In the space homogeneous case the moments of f are constant (because of the conservation equations) and f_m is time independent. Integrating the above inequality delivers the following result

$$\text{(6.2)} \qquad H \geq H_m$$

where H_m is the (constant) value of H corresponding to $f = f_m$. Equation (3.2) says that H is bounded from below (assuming temperature to be finite).

When dealing with the second difficulty ($\dot{H} \to 0$ for $t \to \infty$) then a deeper study of the properties of the Boltzmann equation is needed. In fact, the existence of a limit of H for $t \to \infty$ does not imply that $\dot{H} \to 0$. However, H is not just an arbitrary function of time but a functional of f. Accordingly, although the traditional argument is rather cavalier, objections based on genral remarks about the possible nonexistence of the limits of f and h for $t \to \infty$ are not decisive. In fact the quantity

$$\text{(6.3)} \qquad h = \int (f\log f - f\log f_m + f_m - f)d\underline{\xi} = H - H_m \geq 0$$

is a convex functional of f, vanishing when $f = f_m$ and satisfying $h = S$ where S is definied in Eq. (3.10).

It is tempting to assume that

$$(6.4) \qquad S \equiv \int \log f Q(f,f) d\underline{\xi} \leq -\lambda \rho h$$

where ρ is, as usual, the density and λ a suitable constant. If Eq. (6.4) is assumed to be true, then

$$(6.5) \qquad h \leq -\lambda \rho h$$

and, ρ being constant,

$$(6.6) \qquad h \leq h_0 e^{-\lambda \rho t}$$

Hence $h \to 0$ when $t \to \infty$ and $H \to H_m$. In addition, if f goes to a limit $\bar{f}$ when in a function set where h is a continuous functional, the $\bar{f}$ must be f_m in agreement with the traditional argument.

We examine now the conjecture that Eq. (6.4) holds. The ratio

$$(6.7) \qquad \frac{S}{\rho h} = \frac{\int \log f Q(f,f) d\underline{\xi}}{[\int f d\underline{\xi}] \ \{\int [f \log(f/f_m) + f_m - f] d\underline{\xi}\}}$$

is negative. The ratio is not well defined when $f = f_m$: it seems reasonable to argue that if $S/\rho h \to 0$ for some f, this will happen in neighbourhood of f_m (in a suitable topology).

If this is granted, the one is led to examine the functional

$$(6.8) \qquad J(p) = \frac{2(p, L_m p)}{\rho_m (p,p)_m}$$

where p is the perturbation of f_m(i.e. $f = f_m(1 + p)$), L_m the collision operator linearized about f_m the scalar product notation

$$(6.9) \qquad (p,q)_m = \int f_m(\underline{\xi}) p(\underline{\xi}) q(\underline{\xi}) d\underline{\xi}$$

is used. The functional in Eq. (6.8) is obtained by neglecting terms of order higher than second in both the numerator and the denominator of $S/\rho h$.

If the molecules are rigid spheres on interact with an inverse power law with the exponent $s \geq 5$, then a constant λ_m is known to exist such that $J(p) \leq -\lambda_m$ 25 . This argument is, of course, not a proof of Eq. (6.4), but gives a strong support, in the author's opinion, to the conjecture that Eq. (6.4) holds.

We have to stress, however, that the validity of Eq. (6.4) is not enough to prove the trend to equilibrium since we have to assume that f has a limit $\bar{f}$ when $t \to \infty$.

This is really the basic point of the use of the H-theorem because, if $f \to \bar{f}$ in a function set, where h and S are continuous functionals, then S and h must have a limit; if h has a limit, it must be zero. Hence S also tends to zero; then $\bar{f}$ equals f_m.

The proof of the existence of $\bar{f}$, however, requires an existence proof. This explains why any rigorous use of H-theorem must follow an existence theorem. This kind of result, as explained in the previous section, has been obtained in the space homogeneous case, while its achievement in general appears to be of supreme difficulty.

We maintain that there is no objection to the extension of the consequences of H-theorem in the space inhomogeneous case from a physical point of view, i.e. if we assume that a solution sufficiently smooth exists and we deduce its properties in a formal way. This of course can produce the objections of a critical mathematician who must first prove existence, smoothness, existence of an asymptotic behaviour etc.; it is felt, however, that in the absence of such a refined theory it is important to argue in a formally correct way, leaving the more capable mathematicians the proofs of existence which are required on a stricter level of rigor.

In the space inhomogeneous case, in addition to smoothness, one has to impose an additional condition which is required in order to conclude that H has a limit: the integral in the right hand side of Eq. (3.19) must be nonpositive (no H is supplied to the gas). A sufficient condition is

$$(6.10) \qquad (\underline{q} \cdot \underline{n})_{solid} \leq 0$$

i.e. there is no local influx of energy into the gas. In addition, of course, a costant Maxwellian f_m to which Eq. (3.8) applies must be available.

In the space inhomogeneous case one has also to deal with the circumstance that the density goes to zero; then the H-theorem remains true if the function identical with zero in velocity space is considered to be a form of degenerate Maxwellian. This circumstance appears in a gas expanding in a infinite space with Maxwellian distribution. As noted by PITTERI |33| and quoted by TRUESDELL and MUNCASTER |9| the ratio H/ρ is large than H_m/ρ even in the limit $t \to \infty$ and f/ρ does not tend to a Maxwellian ; H tends to $H_m = 0$, however, and $f \to f_m \equiv 0$ for $t \to \infty$. A zero density is a problem, anyway, as Eq. (6.5) shows.

We also remark that the tendency to equilibrium is not applicable to the homoenergetic solutions to be examined in sect. 6 because there is no constant Maxwellian f_m which can play a role similar to that played by the local Maxwellian in the previous argument. In fact, the temperature is unbounded in time because work is exerted on the gas at a constant rate.

The treatment of the H-theorem in this section has been based, so far, on a paper of the author |34|. Subsequently T. HELMROTH |35| has proved that if a time-dipendent distribution function has a constant mass, momentum and energy and the H-functional tends for $t \to \infty$ to H_m, then f converges strongly to f_m in L^1.

Rather then giving Elmroth's proof here, we indicate a simpler argu ment based on a improvement of inequality (6.1).

In fact it is a simple exercise to prove that this inequality remains true if we add to the right hand side a term R defined as follows:

$$(6.11) \qquad R = cg\left(\frac{|f - f_m|}{f_m}\right) |f - f_m|$$

where c is a numerical constant (indipendent of f) and

$$(6.12) \qquad g(x) = \begin{cases} x & \text{if} \quad 0 \leq x \leq 1 \\ 1 & \text{if} \quad x \geq 1 \end{cases}$$

Integrating then both sides of the modified inequality we obtain

$$(6.13) \qquad H - H_m \geq c\left[\int_{L_t} |f - f_m| d\underline{\xi} + \int_{S_t} \frac{|f - f_m|^2}{f_m} d\underline{\xi} \right]$$

where L_t and S_t denote the sets (depending on time) where f is larger (re spectively, smaller) than f_m. Since H is assumed to tend to f_m it follows that both integrals tend to zero when $t \to \infty$. The fact that the second integral tends to zero implies, by the Schwarz inequality that

$$(6.14) \qquad \int_{S_t} |f - f_m| d\underline{\xi} \to 0$$

then

$$(6.15) \qquad \int |f - f_m| d\underline{\xi} = \int_{L_t} |f - f_m| d\underline{\xi} + \int_{S_t} |f - f_m| d\underline{\xi}$$

also tends to zero and the strong convergence of f to f_m in L^1 is proved.

We remark that this results applies to the space inhomogeneous case as well.

7. EXACT SOLUTIONS IN THE SPACE HOMOGENEOUS CASE

The aim of this section is to study the trend toward equilibrium of a spatially homogeneous gas. This problem is trivial from the fluid dynamic point of view because there is no change at all in density, velocity, temperature and hence no fluid-dynamics. Interesting facts occur in velocity space, however, and the mathematics is by no means trivial. Although Boltzmann's H-theorem guarantees that any assigned initial space independent distribution function will eventually decay to a Maxwellian and this result can be even proved in all rigor, the details of this decay are not known explicitly. An exception is offered by Maxwell's molecules, for which Maxwell |36| derived a set of exact equations satisfied by moments. He did not write out all the terms, which were first published by GRAD |37| and scholarly discussed by IKENBERRY and TRUESDELL |38| and TRUESDELL |39|. In the latter paper, the general solution for the moments of the distribution function in a space homogeneous problem is given by the following expression:

$$N_\alpha = \sum_{i=1}^{N_\alpha} A_{\alpha i}(t) e^{-\lambda_{\alpha i} t} + M_\alpha^{(0)} \tag{7.1}$$

where M_α is any moment and α a suffix to identify it, $M_\alpha^{(0)}$ the equilibrium value of M_α, $A_{\alpha i}(t)$ a polynomial in t, $\lambda_{\alpha i}$ a positive constant, N_α a positive integer; $M_\alpha^{(0)}$, $\lambda_{\alpha i}$ and the coefficients in $A_{\alpha i}(t)$ depend on the (constant) values of density, velocity and temperature. It is not proved by Truesdell's theorem, but only conjectured that the $A_{\alpha i}(t)$ are in fact constants, i.e. zero degree polynomials. The esplicit expressions for $A_{\alpha i}$ can be found by recursion; easy, albeit tedious computations for the first few moments seem to confirm the conjecture.

It is to be remarked that, although a knowledge of all the moments is theoretically equivalent to a knowledge of the distribution function, no simple expression of the latter is obtained from Eq. (4.1). This explains the interest produced by the recent publication |40| of an exact solution of the nonlinear Boltzmann equation for Maxwell's molecules. This solution corresponds to a special initial datum, i.e.

$$f = (b + c\xi^2)\, e^{-a\xi^2} \tag{7.2}$$

where ξ is the molecular speed, a, b, c positive constants. The basic property of a distribution function of the form (7.2) is that it evolves preserving its shape, the only change being in the coefficients a, b, c, which vary in time till, ultimately, c tends to zero, a and b to non zero constants and f to a Maxwellian distribution. It is to be noted that the same solution has been indipendently found by A. V. BOBYLEV |41|.

The starting point of KROOK and WU |40| is the set of the exact moment equations mentioned above, under the assumption of isotropy. In this case f depends on the molecular speed, but not on the direction of molecular velocity and the only moments of interest are

(7.3) $$P_n = \int f(\xi,t)\, \xi^{2n} d\underline{\xi}$$

The equation satisfied by these moments take a verysimple form if we let

(7.4) $$M_n = p_n / [1 \quad 3 \quad 5 \quad 7 \ldots (2n+1)]$$

and normalize M_0 to unity (which is possible since it is constant). Then the moment equations may be reduced to

(7.5) $$\frac{dM_n}{dt} + M_n = \frac{1}{n+1} \sum_{k=0}^{n} M_{n-k} M_k$$

where t is measured in units of a suitable "mean free time". If we further scale speeds with $\sqrt{RT}$, $M_n \to 1$ for $t \to \infty$; in particular $M_1 = 1$.

Krook and Wu re-wrote Eq. (7.5) as an equation for the generating function of moments, defined by:

(7.6) $$M(t,s) = \sum_{n=0}^{\infty} M_n(t) s^n$$

In fact, Eqs. (7.6) and (7.5) give

(7.7) $$\frac{\partial}{\partial s} s \left(\frac{\partial M}{\partial t} + M\right) = M^2$$

and once M is found by solving this equation, the moments M_k are obtained by simply expanding M into a power series in s.

By suitable considerations Krook and Wu are able to determine a solution of Eq. (4.7) which satisfies the conditions:

(7.8) $$M(t,s) = 1 + s + O(s^2) \qquad \text{as} \quad s \to 0$$
$$M(t,s) \quad \frac{1}{1-s} \qquad \text{as} \quad t \to \infty$$

as required by the normalisation of the moments. The solution is as follows:

(7.9) $$M(t,s) = \frac{1 + [1 - 2K(t)]s}{[1 - K(t)s^{\,2}]}$$

where

(7.10) $$K(t) = 1 - \exp(-\frac{t}{6})$$

provided a suitable origin is chosen for times.

The normalized moments are then given by

(7.11) $$M_n(t) = K^{n-1}[n - (n-1)K]$$

The distribution function is obtained by a Fourier transform technique and turns out to be

(7.12) $$f(\xi,t) = (2\pi)^{-3/2}\left(\frac{5K-3}{2K^{5/2}} + \frac{1-K}{2K^{7/2}}\xi^2\right)\exp\left(-\frac{\xi^2}{2K}\right)$$

Once one is convinced that the Boltzmann equation for Maxwell's molecules has a solution of the form

(7.13) $$f = e^{-a\xi^2}(b + c\xi^2)$$

with $a = a(t)$, $b = b(t)$, $c = c(t)$ (a fact which was pointed out to me in 1967 by ROY KRUPP |42|, a student then working on his master's thesis at M.I.T.) it is not complicated to find a, b and c. As a matter of fact, if f is given by Eq. (4.13), one can easly compute the collision term:

(7.147 $$Q(f,f) = Q(c\xi^2 e^{-a\xi^2}, c\xi^2 e^{-a\xi^2}) = -\tfrac{1}{2}c^2 e^{-a\xi^2} L(\xi^4)$$

where L is the collision operator linarized about the Maxwellian $f_0 = e^{-a\xi^2}$ (Here use has been made of the fact

(7.15) $$\xi'^2 + \xi'^2_* = \xi^2 + \xi_*^2$$

and hence

(7.14) $$\xi'^4 + \xi'^4_* - \xi^4 - \xi_*^4 = -2(\xi'^2\xi'^2_* - \xi^2\xi_*^2)$$

if $\underline{\xi}$, $\underline{\xi}_*$, $\underline{\xi}'$, $\underline{\xi}'_*$ denote the velocities of two molecules before and after collision). $L(\xi^4)$ is a second degree polynomial for Maxwell molecules:

(7.17) $$L(\xi^4) = -2K(4\xi^4 - 2a^{-1}\xi^2 + 15a^{-2})a^{-3/2}$$

where K is a positive constant. Hence the Boltzmann equation to be satisfied becomes:

(7.18) $$\left[-\frac{da}{dt}\xi^2(b + c\xi^2) + a\frac{db}{dt} + a\frac{dc}{dt}\xi^2\right]e^{-a\xi^2} = e^{-a\xi^2}Kc^2a^{-3/2}(4\xi^4 - 20a^{-1}\xi^2 + 15a^{-2})$$

This equation is satisfied if and only if

$$-\frac{da}{dt} = 4Kca^{-3/2}$$

$$\text{(7.19)} \qquad -\frac{da}{dt}\, b + \frac{dc}{dt} = -20ka^{-5/2}\, c^2$$

$$\frac{db}{dt} = 15\ k\ a^{-7/2}\ c^2$$

Elimination of the time variable gives an easily integrated linear system for b and c as functions of a; the following result is obtained

$$\text{(7.20)} \qquad c = A \quad (2a)^{7/2} - (2a)^{5/2} \qquad b = -3A(2a)^{5/2} + 5A(2a)^{3/2}$$

where A is a constant which has to be taken equal to $\frac{1}{2}(2\pi)^{-3/2}$ in order to comply with normalization. Eq. (7.12) is thus recovered provided 2a is identified with K^{-1}. The first equation of system (7.19) may then be used to find a = a(t), after re-writing it in the following form

$$\text{(7.21)} \qquad -\frac{da}{dt} = 4kA \quad (2a)^2 - 2a$$

Hence

$$\text{(7.22)} \qquad K = (2a)^{-1} = 1 - B\ e^{-8kAt}$$

which coincides with Eq. (7.10) provided the same normalizing conditions are assumed $(B = 1;\ k = (48\ A)^{-1})$.

The BKW solution (called sometimes in the literature also BKW-mode) approaches an equilibrium distribution when $t \to \infty$ in a nonuniform fashion; this is due to the high speed tail of the distribution and indicates that linearization does not hold for high sppeds even when we are close to a Maxwellian in some sense. This was clear from other facts (loss of positivity of linearized solutions) but is immediately obvious here. Physically, as remarked by Krook and Wu, this is due to the fact that, at most, the total kinetic energy of two molecules after a collision can be concentrated in one of them. If the tail is initially absent above a certain energy (cutoff energy), then this value can at most double after each favorable collision. Thus, the time required to reach a higher cutoff can be expected to grow logarithmically with the latter.

The importance of the BKW solution was overestimated initially, because of the following conjecture formulated by Krook and Wu: an arbitrary initial state tends first to relax towards a BKW mode; then a relaxation according to the latter takes place. This conjecture can rephrased in more mathematical terms, but this is not necessary, since both numerical and analytical evidence against this conjecture have been found by many authors.

An enormous literature on this subject and related solutions of other models than Maxwell is available. For a survey, we refer to the papers by M. H. Ernst |43, 44|.

8. INTRODUCTION TO AFFINE HOMOENERGETIC FLOWS

At about the same time (1956) C. TRUESDELL |45| and V.S. GALKIN |46| independently investigated the steady homoenergetic flows of a gas of Maxwellian molecules according to the infinite system of moments associated with the Boltzmann equation. Later Galkin |47 - 49| extended his analyses to some typical unsteady homoenergetic affine flows. These analyses are discussed and summarized in the book by Truesdell and Muncaster on kinetic theory |50|.

While these analyses have the great advantage of leading to explicit solutions, which lend themeselves to a detailed discussion of their properties, they suffer of two drawbacks:

1) They are restricted to Maxwellian molecules
2) They provide solutions to the system of moments,but no proof is given of the existence of a corresponding solution of the Boltzmann equation itself.

The aim of this part is to complement the abovementioned analyses by proving an existence theorem for the Boltzmann equation and general molecular models, when the initial data are compatible with a homoenergetic affine flow. Some of the data lead to an implosion and infinite density in a finite time, in agreement with the physical picture of the associated flows; for the remaining set of data, global existence will be shown to hold.

It is to be remarked that the teorem to be proved appears to be the only one available for the space inhomogeneous Boltzmann equation without restriction on the size of the data.

The treatment follows a recent paper by the author |51|. To start with,it is convenient to recall the basic ideas about homoenergetic affine flows. The defining properties are the following ones:

a) The body force (per unit mass) $\underline{X}$ acting on the molecules is constant:

(8.1) $$\underline{X} = \text{constant}$$

b) The density ρ, the internal energy per unit mass e, the stress tensor $\underline{p}$ and the heat flux $\underline{q}$ may be functions of time but not of the space coordinates.

c) The bulk velocity $\underline{v}$ is an affine function of position $\underline{x}$:

(8.2) $$\underline{v} = \underline{\underline{K}}(t)\underline{x} + \underline{v}_0(t)$$

This definition holds for a general material; for a gas described by kinetic theory, a natural extension of property b) is spontaneous:

b') The moment formed with the peculiar velocity:

(8.3) $\underline{c} = \underline{\xi} - \underline{v}$

may be the functions of time but not of space coordinates. Here $\underline{\xi}$ is the molecular velocity with respect to an inertial frame.

We remark that b') holds for the solutions obtained by Truesdell |45| and Galkin |46 - 49|. When we work with the distribution function f, this condition transform itself into

b') The variable $\underline{x}$ appears in f only through $\underline{v}$, given by Eq. (8.2), i.e.:

(8.4) $f = f(\underline{c},t)$

An analysis of the balance equations based on a), b), c) immediately leads to the following restrictions on $\underline{\underline{K}}$ and $\underline{v}_0$:

$$\dot{\underline{\underline{K}}} + \underline{\underline{K}}^2 = 0$$

(8.5)

$$\dot{\underline{v}}_0 + \underline{\underline{K}}\underline{v}_0 = \underline{X}$$

The general solution of this system is

$$\underline{\underline{K}}(t) = \left[\underline{\underline{I}} + t\underline{\underline{K}}(0)\right]^{-1}\underline{\underline{K}}(0)$$

(8.6)

$$\underline{v}_0(t) = \left[\underline{\underline{I}} + t\underline{\underline{K}}(0)\right]^{-1}\left[\underline{v}_0(0) + t\underline{X} + \tfrac{1}{2}t^2\underline{\underline{K}}(0)\underline{X}\right]$$

where $\underline{\underline{I}}$ is the 3x3 identity matrix. This solution exists globally for $t \to 0$ if the eigenvalues of $\underline{\underline{K}}(0)$ are nonnegative; otherwise the solution ceases to exist for $t = t_0$, where $-t_0^{-1}$ is the largest, in absolute value, among the negative eigenvalues of $\underline{\underline{K}}(0)$.

In particular if,

(8.7) $\left[\underline{\underline{K}}(0)\right]^2 = 0$

then $\underline{\underline{K}}(t)$ is indipendent of time, because $\left[\underline{\underline{I}} + t\underline{\underline{K}}(0)\right]^{-1} = \underline{\underline{I}} - t\underline{\underline{K}}(0)$.

$\underline{v}$ is then steady if and only if

(8.8) $\underline{\underline{K}}(0)\underline{X} = 0$

and $\underline{v}_0(0)$ is chosen in such a way that

(8.9) $\underline{\underline{K}}(0)\underline{v}_0(0) = \underline{X}$

In particular this is always possible if $\underline{X} = 0$.

Eq. (2.7) is satisfied if and only if a coordinate system exists for which the matrix associated by $\underline{\underline{K}}(0)$ is given by

(8.10) $$((K_{ij})) = \begin{Vmatrix} 0 & 0 & 0 \\ k & 0 & 0 \\ 0 & 0 & 0 \end{Vmatrix}$$

For a simple proof of this see the Appendix of Ref. |51|.

Eqs. (8.5) are certainly necessary for a solution satisfying conditions a), b''), c) to exist since they are derived (6) under the assumption a), b), c) and b'') implies b).

In order to show that Eqs (2.5) are also sufficient, we consider the Boltzmann equation:

(8.11) $$\frac{\partial f}{\partial t} + \underline{\xi} \cdot \frac{\partial f}{\partial \underline{x}} + \underline{X} \cdot \frac{\partial f}{\partial \underline{\xi}} = Q(f,f)$$

where Q(f,f) is the collision operator | 2 - 5 |. We choose $\underline{c}$ in place of $\underline{\xi}$ as independent variable and use the same letter f for $f(\underline{x}, \underline{\xi}, t)$ and $f(\underline{c}, t)$ although, of course, they are different functions of their arguments. Then if Eq. (8.4) holds we have to make the following replacements in Eq. (8.11)

(8.12)
$$\frac{\partial f}{\partial t} \rightarrow \frac{\partial f}{\partial t} - \underline{\xi}\,\frac{\partial f}{\partial c} \cdot \underline{\underline{K}}x - \frac{\partial f}{\partial \underline{c}} \cdot \underline{v}_0$$
$$\frac{\partial f}{\partial \underline{x}} \rightarrow -\frac{\partial f}{\partial \underline{c}} \cdot \underline{\underline{K}}$$

Eq. (8.11) then becomes:

(8.13) $$\frac{\partial f}{\partial t} - \frac{\partial f}{\partial \underline{x}} \cdot (\underline{\underline{K}} + \underline{\underline{K}}^2)\underline{x} - \frac{\partial f}{\partial \underline{c}} \cdot (\underline{v}_0 + \underline{\underline{K}}\underline{v}_0 - \underline{X}) - \frac{\partial f}{\partial \underline{c}} \cdot \underline{\underline{K}}\underline{c} = Q(f,f)$$

where Q(f,f) is now expressed in terms of $\underline{c}$ rather than $\underline{v}$.

In order for a solution independent of $\underline{x}$ to exist, as required by Eq. (8.4), the first equation of system (8.5) must be satisfied. If the second is also fullfilled, then Eq. (8.13) becomes:

(8.14) $$\frac{\partial f}{\partial t} - \frac{\partial f}{\partial \underline{c}} \cdot \underline{\underline{K}}\underline{c} = Q(f,f)$$

and the space variable no longer appears explicitly. It is to be emphasized that the second equation of system (8.5) is also necessary for a solution satisfying Eq. (8.4) to exist, because multiplying Eq. (8.13) by $\underline{c}$ and integrating yields exactly that equation, provided one recalls that, by definition:

(8.15) $$\int \underline{c} f d\underline{c} = 0$$

Thus the existence of homoenergetic affine flows is reduced to proving an existence theorem for Eq. (8.13). We notice that the latter can be cast into an integral form provided we determine the semigroup corresponding to collisionless flow.

To this end we consider the ordinary differential equation:

(8.16) $$\frac{d\underline{c}}{dt} = -\underline{\underline{K}}\underline{c} \qquad \underline{c}(0) = \underline{c}_0$$

which can be easily solved if the expression of $\underline{\underline{K}}$ appearing in Eq. (8.6) is used. We obtain

(8.17) $$\underline{c} = \left[1 + t\underline{\underline{K}}(0)\right]^{-1}\underline{c}_0$$

Hence if we let

(8.18) $$f^*(\underline{c},t) = f(\left[1 + t\underline{\underline{K}}(0)\right]^{-1}\underline{c},t)$$

we obtain

(8.19) $$\frac{\partial f^*}{\partial t} = \left(\frac{\partial f}{\partial t} - \frac{\partial f}{\partial \underline{c}} \cdot \underline{\underline{K}}\underline{c}\right)^* = \left[Q(f,f)\right]^*$$

Integration with respect to t yields the integral form

(8.20) $$f^*(\underline{c},t) = f^*(\underline{c},0) + \int_0^t \left[Q(f,f)\right]^* (\underline{c},s)ds$$

Other integral forms are possible when Q(f,f) can be split into two separate contributions (gain and loss terms), as is the case for hard sphere molecules and cutoff interactions.

9. A PRIORI ESTIMATES OF MASS AND ENERGY DENSITY

In order to obtain an existence theorem, we need a priori estimates on the mass and energy densities, i.e. on the moments

(9.1) $$\rho = \int f d\underline{c} \qquad 2E = \int c^2 f d\underline{c} = \mathrm{Tr}\underline{\underline{p}}$$

where Tr denotes the trace of a tensor.
Eq. (8.14) yields the following equations

(9.2) $$\frac{d\rho}{d\rho} + \rho \,\mathrm{Tr}\, \underline{\underline{K}} = 0$$

(9.3) $$\frac{dE}{dt} + \mathrm{Tr}(\underline{\underline{K}}\underline{\underline{p}}) + 2E\,\mathrm{Tr}\,\underline{\underline{K}} = 0$$

Eq. (9.2) immediately yields

(9.4) $$\rho(t) = \rho(0)\exp\left(-\int_0^t \mathrm{Tr}\,\underline{\underline{K}}(s)ds\right)$$

and since $\underline{\underline{K}}(s)$ is explicitly known, the density is determined. Let now k denote the largest absolute value of the elements of the matrix $\underline{\underline{K}}$. Then, if we use the inequality

(9.5) $$|p_{ij}| \le \tfrac{1}{2}(p_{ii} + p_{jj})$$

Eq. (9.3) gives

(9.6) $$\frac{dE}{dt} \le 3kE - 2E\,\mathrm{Tr}\,\underline{\underline{K}}$$

and

$$E(t) \le E(0)\exp\left[3\int_0^t k(s)ds - \int_0^t \mathrm{Tr}\,\underline{\underline{K}}(s)ds\right]$$

We remark that $|\mathrm{Tr}\,\underline{\underline{K}}|$ and k are bounded for $t \le \bar{t} < t$ where $-t_0^{-1}$ is the largest among the negative eigenvalues of $\underline{\underline{K}}(0)$. If there are no negative eigenvalues then k, $|\mathrm{Tr}\,\underline{\underline{K}}|$ are bounded for any positive t.

10. THE EXISTENCE THEOREM

In oreder to prove the existence of a solution for cutoff collision terms we first consider a collision term

$$Q(f,f) = \iiint B(\theta, |\underline{c} - \underline{c}_*|)(f'f_*' - ff_*)d\underline{c}_* d\theta de \tag{10.1}$$

where the kernel $B(\theta, |\underline{c} - \underline{c}_*|)$ is bounded:

$$B(\theta, |\underline{c} - \underline{c}_*|) \leq \frac{a}{\pi^2} \qquad (a = \text{constant}) \tag{10.2}$$

Then

$$J(f,f) = Q(f,f) + a f \int f_* d\underline{c}_* \tag{10.3}$$

is a positive functional of the distribution function and increases when f increases. If $\phi(\underline{c})$, the initial value of f, is in L^1, let us consider the iteration scheme where $\rho(t)$ is given by Eq. (9.4)

$$\begin{aligned} &\frac{\partial f^{\#}_{n+1}}{\partial t} + a\rho(t)f^{\#}_{n+1} = J(f_n, f_n)^{\#} \qquad f_{n+1}(\underline{c},0) = \phi(\underline{c}) \quad (n \geq 0) \\ &f_0(\underline{c},t) = 0 \end{aligned} \tag{10.4}$$

Here t runs from 0 to T, where T is arbitrary if $\rho(t)$ exists for any positive t, less than t_0 if $\rho(t)$ exists finite only for $t < t_0$.

Then f_n is a monotonously increasing sequence whose norm in L^1 is bounded by $\rho(t)$; hence it tends to an L^1 function $f(\underline{c},t)$. The latter will have a density $\hat{\rho}(t)$ such that

$$\begin{aligned} &\frac{d\hat{\rho}}{dt} + a\hat{\rho}\mathrm{Tr}\underline{\underline{K}} = a\hat{\rho}(\hat{\rho} - \rho(t)) \\ &\hat{\rho}(0) = \rho(0) \end{aligned} \tag{10.5}$$

The unique solution solution of this initial value problem is

$$\hat{\rho} = \rho(0) \tag{10.6}$$

Hence f will satisfy

$$\begin{aligned} &\frac{\partial f^{\#}}{\partial t} = Q(f,f)^{\#} \\ &f(\underline{c},0) = \phi(\underline{c}) \end{aligned} \tag{10.7}$$

Hence a solution in L exists. This solution is the unique solution of the initial value problem (10.7); it is also unique among the solutions with the same initial value and bounded density.

If we assume that the initial data have a finite second moment E(0), then the inequality for E(0) discussed in the previous section gives

(10.8) $$E(t) \leq A_T = E(0)\exp(3\int_0^T k(s)ds) \qquad (0 \leq t \leq T)$$

We remark that the constant A_T depends on the initial values only (because so does k(t) and T.

We can now prove a rigorous H-theorem for Eq. (10.7), under the assumption that the initial data have a finite H-functional, H(0), where, in general

(10.9) $$H(t) = \int f\log f d\underline{c}$$

Formally, remarking that:

(10.10) $$\mathrm{Det}(1+t\underline{\underline{K}}(0)) = e^{\int_0^t \mathrm{Tr}\underline{\underline{K}}(s)ds} \equiv D(t)$$

we have

(10.11) $$\frac{d}{dt} D(t)H(t) \leq 0$$

and hence

(10.12) $$H(t) \leq H(0)/D(t)$$

In order to justify these formal steps we use the method of ARKERYD |31| and introduce the modified initial data

(10.13) $$\phi_{n,p}(\underline{c}) = \min\left[\phi(\underline{c}) + \frac{1}{n}e^{-c^2}, p\right]$$

which tend to ϕ when $n \to \infty$, $p \to \infty$. With the help of a suitable cutoff of the collision term we arrive at Eq. (10.2) rigorously as in Ref. |31|, where the factor D(t) was, of course, missing.

As a consequence

(10.14) $$H(t) \leq H_T$$

where, again, the constant H_T depends only on the initial data and T.

When the boundedness assumption stated in Eq. (10.2) does not hold, we can consider the more general condition:

(10.15) $$B(\theta, |\underline{c} - \underline{c}_*|) \leq \frac{b}{\pi^2}(1 + c^2 + c^2) \qquad (b = \text{constant})$$

To deal with this case we first replace b by a cutoff expression

(10.16) $$B_m(\theta, |\underline{c} - \underline{c}_*|) = \min(B(\theta, |\underline{c} - \underline{c}_*|), m)$$

where m is a positive constant. Then

(10.17) $$B_m(\theta, |\underline{c} - \underline{c}_*|) \leq a_m/\pi^2 \qquad (a_m = \text{constant})$$

and we can apply the previous result to conclude that there is a solution f_m of

(10.18) $$\frac{\delta f_m^*}{\delta t} = Q_m(f_m, f_m) *$$

$$f_m(\underline{c},0) = \phi(\underline{c})$$

In addition if the inital values possess a finite moment E (0) and a finite H-functional, H(0), these functionals will exist at any time $t \leq T$ and will satisfy the inequalities:

(10.19) $$E_m(t) \leq E_T$$

(10.20) $$H_m(t) \leq H_T$$

where the constants E_T and H_T do not depend on m. Then using the same weak compactness criterion used by ARKERYD |32| in his first proof and and the equicontinuity in time of sequence f_m, we arrive at the following.

Existence theorem: there exists a solution f of Eq. (10.7), where the kernel $B(\theta, |\underline{c} - \underline{c}_*|)$ of the collision term Q(f, f) satisfies Eq. (10.15)and the intial density, energy density and H-functional are finite at time 0. These functional remains bounded for $0 \leq t \leq T$.

If we add the assumption that the fourth moment of ϕ exists, then we can prove a uniqueness theorem. To this end one has to prove that the fourth moment remains finite:

(10.21) $$Q(t) \equiv \int (1 + c^2)^2 f(\underline{c},t) d\underline{c} \leq C_T \quad (0 \leq t \leq T)$$

This is easly done by using a special case of POVZNER inequality |52|:

(10.22) $$(1+c'^2)^2 + (1+c'^2_*)^2 - (1+c^2)^2 - (1+c_*^2) \leq 2(1+c^2)(1+c_*^2)$$

to conclude that

(10.23) $$\int (1+c^2)^2 \, Q(f,f) d\underline{c} \leq 4b(E_T + R_T)Q(t)$$

where R_T is the maximum value of the density in |0,T|. Hence Eq. (4.7) gives:

(10.24) $$\frac{dQ}{dt} + Q\mathrm{Tr}\underline{\underline{K}} + 4\int \underline{c}\cdot\underline{\underline{K}}\underline{c}(1+c \;) f d\underline{c} \leq 4b(E_T + R_T)Q$$

and Eq.(4.21) follows with:

(10.25) $$C_T = Q(0) \exp\{|4b(E_T + R_T) + 12K + M_T \mid T\}$$

where M_T is the maximum value of $\mathrm{Tr}\underline{\underline{K}}$ 0,T.

It is now easy to prove the following

Uniqueness theorem. Let f be the solution, whose existence is guaranteed by the previous theorem. If Q(0) exists, then so does Q(T) in |0,T| and the solution of Eq. (10.7) is unique among those having this property.

REFERENCES

1) C. CERCIGNANI - Transport Theory and Statistical Physics, 2, 211 (1982)

2) C. CERCIGNANI - Theory and Application of the Boltzmann Equation, Scottish Academic Press, Edinburgh, and Elsevier, New York (1975)

3) M. AIZENMAN - Duke Math. J., 45? 809 (1978)

4) H. GRAD - In Handbuch der Physik, Vol. XII, Springer, Berlin (1958)

5) C. CERCIGNANI - Mathematical Methods in Kinetic Theory, Plenum Press, New York, and Mc Millan, London (1969)

6) O. LANFORD - In Springer Lecture Notes in Physics, 38, 1, ed. J. Moser, Springer, Berlin (1975)

7) H. SPOHN - In "Kinetic Theories and Boltzmann Equation", ed. C. Cercignani, LNM 1048, Springer (1984).

8) C. CERCIGNANI - Transport Theory Statistical Physics, 2 (1972) 27.

9) C. TRUESDELL and R.G. MUNCASTER - Fundamentals of Maxwell's Kinetic Theory of a Simple Monatomic Gas (Academic Press, New York, 1980)

10) C. CERCIGNANI and M. LAMPIS - Transport Theory Statistical Physics, 1(1971) 101

11) J. DARROZES and J. GUIRAUD - Compt. Rend. Acad. Sci. (Paris) A262 (1966) 1368

12) J. SCHNUTE - Canad. Jour. Math., 27, 1271, 1975

13) L. BOLTZMANN - Sitzungsberichte der Akademie der Wissenschaften, Wien, 66, 275; Wissenschaftliche Abhandlungen, 1, 316, 1872

14) L. BOLTZMANN - Sitzungsberichte der Akademie der Wissenschaften, Wien, 72, 427; Wissenschaftliche

15) H.A. LORENZ - Wien Ber. 95: 115 (1887) {also in Collected Papers, Martinus Nijhoff, ed., Vol. 6 p. 74}

16) S. CHAPMAN and T. G. COWLING - The Mathematical Theory of Non-Uniform gases (Cambridge University Press. Cambridge, 1960)

36) J.C. MAXWELL - "Scientific Papers" 2, 26 and 2, 680, Dover, New York (1965)

37) H. GRAD - Comm. Pure Appl. Math., 3, 331 (1949)

38) E. IKENBERRY and C. TRUESDELL - J. Rat. Mech. and Analysis, 5, 1 (1956)

39) C. TRUESDELL - J. Rat. Mech. and Analysis, 5, 55 (1958)

40) M. KROOK and T.T. Wu - Phys. Rev. Lett. 36, 1107 (1976) and Phys. Fluids 20, 1589 (1977)

41) A.V. BOBYLEV - Sov. Phys. Doklady, 20, 820 and 822 (1976) and 21, 632 (1977)

42) R. KRUPP - Private comunication (1967)

43) M. H. ERNST - Phys. Rep. 78, 1 (1981)

44) M. H. ERNST - In Nonequilibrium Phenomena I: The Boltzmann Equation, J.L. LEBOWITZ and E. W. MONTROLL, eds. North Holland, Amsterdam (1983)

45) C. TREUSDELL - "On the pressures and the flux of energy in a gas according to Maxwell's Kinetic theory, II", Journal of Rational Mechanics and Analysis, 5, 55-128 (1956)

46) V. S. GALKIN - Prikladnaya Matematika i Mechanika, 20, 445-446 (1956) (In Russian)

47) V. S. GALKIN - "On a class of solutions of Grad's moment equations", PMM 22, 532- 536 (1958)

48) V. S. GALKIN - "One-dimensional unsteady solution of the equation for the kinetic moments of a monatomic gas", PMM 28, 226-229 (1964)

49) V. S. GALKIN - "Exact solutions of the kinetic moment equations of a mixture of monatomic gases", Fluid Dynamics, 1, 29-34 (1966)

50) C. TRUESDELL and R. G. MUNCASTER - Fundamentals of Maxwell's Kinetic Theory of a Simple Monatomic Gas, New York, Academic Press (1980)

51) C. CERCIGNANI - Submitted to Archive for Rational Mechanics and Analysis (1986)

17) L. BOLTZMANN - Vorlesungen über Gastheorie, J. A. Barth, ed. (Leipzig, 1898) Vol. 2, Chap VII {also in Lectures on Gas Theory, transl. S.G. Brush (Berkeley, 1964)}

18) R.C. TOLMAN - The Principles of Statistical Mechanics (Oxford University Press, London, 1938)

19) G. H. BRYAN - Brit. Assoc. Reports, p. 64 (1894)

20) F. B. PIDDUCK - Proc. R. Soc. London Ser. A 101 (1922)

21) L. BOLTZMANN - Wien Ber. 95: 153 (1887) {also in Wiss. Abh. 3: 272}

22) G. E. UHLENBECK - In The Boltzmann Equation: Theory and Applications, E. G. D. COHEN and W. THIRRING, eds. Acta Phys. Austr., Suppl X, 107 (Springer-Verlag, Vienna, 1973)

23) L. WALDMANN - Z. Naturforsch, 12A: 660 (1957)

24) L. WALDMANN - Z. Naturforsch, 13A: 609 (1958)

25) R.F. SNIDER - J. Chem. Phys. 32: 1051 (1960)

26) E. C. G. STUECKELBERG - Helv.Pys. Acta 25: 577 (1952)

27) L. WALDMANN - In Handbuch der Physik, S. FLÜGGE ed. (Springer-Verlag, Berlin, 1958), Vol. 12, p. 484

28) C. CERCIGNANI and M. LAMPIS - J. Stat. Phys., 26, 795 (1981)

29) L. WALDMANN - In the Boltzmann Equation: Theory and Applications, E. G. D. COHEN and W. THIRRING, eds. Acta Phys. Austr., Suppl. X, 107, (Springer-Verlag, Vienna 1973)

30) E. WILD - Proc. Camb. Phil. Soc., 47, 602, 1950

31) D. MORGENSTERN - J. Rat. Mech. Anal. 4, 533, 1955

32) L. ARKERYD - Arkive Rat. Mech Anal; 45,1 and 17, 1972

33) M. PITTERI - Acc. Naz Lincei (Roma). Sez. VIII, 67, 248, 1979

34) C. CERCIGNANI- Arch. Mech. 34, 231 (1982)

35) T. ELMROTH - Ph. D. Thesis, Göteborg (1984)

52) A. YA. POVZNER - "The Boltzmann equation in the kinetic theory of gases", American Mathematical Society Translations (2). 47, 193- 216

THE VLASOV EQUATION: SOME MATHEMATICAL ASPECTS

H. Babovsky, H. Neunzert
University of Kaiserslautern, Kaiserslautern, F.R.G.

1. Introduction

The subject of these lectures is the Vlasov-Equation: Its derivability, some properties of (stationary) solutions, and the application in a field quite different from gas dynamics: In nuclear physics.

Since most of the other lectures are concerned with the Boltzmann equation, we should point out at the beginning some of the differences between the Boltzmann equation and the Vlasov equation.

The Boltzmann equation describes a system of gas particles the interaction of which is restricted to short ranges. Therefore, the interaction can be modelled appropriately by collisions. In contrast to this, the Vlasov equation is related to gases with long range interactions, such as plasmas (electromagnetic field) or stellar systems (gravitational field). From these differences result the different shapes of the Boltzmann and the Vlasov equation: The Boltzmann equation reads

$$\left(\frac{\partial}{\partial t} + v \cdot \nabla_x\right) f = J(f,f),$$

where $f = f(t,x,v)$ is the particle distribution in phase space, (x,v) is the position - velocity vector, and $J(f,f)$ models the effect of particle collisions. In contrast to this, the Vlasov equation does not have any collision term, i.e. the right hand side is zero. Particle interactions are described on the side of differentials, thus the left hand side is more complicated. A consequence of this is that the Vlasov equation is reversible while the Boltzmann equation is not.

2. On the Derivation of the Vlasov Equation

The first question to investigate is the following: Can the Vlasov equation be rigorously derived? More precisely: Is the Vlasov equation obtained in the limit of an N-particle system, when the number N of particles goes to infinity?

The starting point of our investigations are particle flows:

2.1 Particle flows

Let $\Gamma = \Omega \times \mathbb{R}^k$ be some given phase space. Here $\Omega \subset \mathbb{R}^l$, $l \leq k$, is the position space while $\mathbb{R}^k$ is the set of velocities which can be obtained by the particles. For the moment, the special form of Γ is of no importance. We define a particle flow on Γ in the most abstract manner as a two-parameter family of measurable bijections

$$T_{t,s} : \Gamma \longrightarrow \Gamma, \quad t,s \in \mathbb{R}$$

with the properties: for all $r,s,t \in \mathbb{R}$,

$$T_{t,s} \circ T_{s,r} = T_{t,r}$$

$$T_{t,t} = \text{id}.$$

(From this follows immediately

$$T_{t,s}^{-1} = T_{s,t} \text{ .)}$$

The interpretation of $T_{t,s}$ is as follows: Given the position $P(s) \in \Gamma$ of a particle in phase space at time s, then $P(t) := T_{t,s} P$ describes the position at time t. Thus, $T_{t,s}$, $t \geq s$, may be interpreted as some dynamics in Γ.

If now a measure μ_{t_o} on Γ is given at time t_o, then a flow of measures may be defined through T_{t,t_o} by

$$\mu_t(T_{t,t_o}A) := \mu_{t_o}(A).$$

Since $T_{t_o,t}$ is the inverse of T_{t,t_o}, this is equivalent to

$$\mu_t(A) = \mu_{t_o}(T_{t_o,t}A) = \mu_{t_o} \circ T_{t_o,t}(A),$$

so that the flow is defined by the "continuity equation"

$$\mu_t = \mu_{t_o} \circ T_{t_o,t}.$$

A first question which is of importance, especially for the derivation of the Vlasov equation, is: Does μ_t depend continuously on t and on the "initial distribution" μ_{t_o}? In general, the answer is certainly: no, since until now we did not demand any continuity property of T_{t,t_o}. However, if T_{t,t_o} is continuous, then the answer is yes, as follows from

Theorem 1: Suppose $(T_n)_{n\in N}$ and T are measurable mappings on Γ; further,

$$\mu_n \longrightarrow \mu \text{ weakly}$$

and

$$\mu\big(\{P\in\Gamma \mid T_n \text{ converges continuously to } T \text{ in } P\}\big) = 1.$$

Then $\mu_n \circ T_n^{-1} \longrightarrow \mu \circ T^{-1}$ weakly.

A proof of this theorem may be found in Billingsley[1].

Since our aim is the Vlasov equation, we are mainly interested in differentiable particle flows defined by velocity fields. We shall explain now what is meant by this.

2.2 Velocity fields

For the moment, we restrict to the case $\Omega = \mathbb{R}^1$. The flows we are going to concider are defined by differential equations. This means, that $P(t) = T_{t,s}P_o$ is a solution of an initial value problem

$$\dot{P}(t) = V(t,P), \quad P(s) = P_o,$$

where V is some given vector field. If the initial value problem is uniquely solvable, then we say that $T_{t,s}$ is a differentiable flow generated by the vector field V.

Now suppose that we start with an initial distribution μ_{t_o} which is absolutely continuous with density f_{t_o}. Are the measures

$$\mu_t = \mu_{t_o} \circ T_{t_o,t}$$

also absolutely continuous? And if they are, how do the densities evolve? The answers are given by the following "General transport theorem":

Theorem 2: If V is locally Lipschitz, then the family $\{\mu_t\}$ of measures defined by the continuity equation has densities f(t) which are weak solutions of

$$\frac{\partial}{\partial t} f + \mathrm{div}\,(f \cdot V) = 0, \quad f(t_o) = f_{t_o}.$$

Proof: A theorem by Rademacher[2] says:

Suppose $T : \Gamma \longrightarrow \Gamma$ is bijective and measurable, and T^{-1} is locally Lipschitz continuous. Then $\mu \circ T^{-1}$ is absolutely continuous if μ is.

Now it is well known from the theory of ODE that $T_{t,s}$ is locally Lipschitz continuous if V is. With $T := T_{t,t_o}$ follows the absolute continuity of μ_t : (λ = Lebesgue measure)

$$\mu_t(M) = \int_M f_t d\lambda = \int_{T_{t_o,t}M} f_{t_o} d\lambda$$

From the transformation rule for integrals follows

$$f_t(P) = f_{t_o}(T_{t_o,t}P) \cdot \Delta_t(P),$$

where

$$\Delta_t = \left| \det \frac{\partial T_{t_o,t}P}{\partial P} \right|$$

is the Jacobian of $T_{t_o,t}$ (if this derivative exists).
With

$$h(t,P) := \frac{\partial \phi}{\partial t}(t,P) + V(t,P)\cdot\nabla_P\phi(t,P)$$

we get therefore, by a straightforward calculation,

$$\int_{t_o}^{t} \left(\int_\Gamma f_s\cdot h\, d\lambda\right) ds = - \int_\Gamma f_{t_o}\, \phi(o,.)\, d\lambda,$$

which is the weak version of the equation of Theorem 2.

<u>Remarks</u>: 1) Suppose $V \in C^{(1)}$ (from which follows that

$$\Delta_t = \left| \det \frac{\partial T_{t_o,t}P}{\partial P} \right| \quad \text{exists).}$$

Then Liouville's theorem tells us that $\Delta_t = 1$ if $\mathrm{div}_P\, V = 0$, and therefore f = const along trajectories:

$$f_t(P) = f_{t_o}(T_{t_o,t}P).$$

2) For a particle flow moving in an electromagnetic field (E,B), the velocity field at $P = (x,v) \in \mathbb{R}^k \times \mathbb{R}^k$ is given by

$$V(t,x,v) = \begin{pmatrix} v \\ \frac{q}{m}\left(E + \frac{1}{c}\, v\times B\right) \end{pmatrix}.$$

Since E and B depend only on t and x, we have

$$\mathrm{div}_P\, V = 0,$$

and the continuity equation reads

$$\frac{\partial f_t}{\partial t} + v\cdot\nabla_x f_t + \frac{q}{m}\left(E + \frac{1}{c}\, v\times B\right)\cdot\nabla_v f_t = 0.$$

If E and B are given, then this equation may be solved by

$$f_t(t,P) = f_{t_o}(T_{t_o,t}\, P),$$

where $T_{t,t_o}P$ is a solution of

$$\dot{x} = v$$

$$\dot{v} = \frac{q}{m}\left(E + \frac{1}{c}\, v\times B\right), \quad \begin{pmatrix} x \\ v \end{pmatrix}(t_o) = P.$$

More complicated is the situation, if E and B depend on the distribution $f_t d\lambda$. (This case is described by the Vlasov equation.) Before we study this case, let's have a short look on boundary conditions.

2.3 Boundary conditions

Now we assume that Ω is a connected region with smooth boundary $\partial\Omega$. In this case, some of the trajectories $T_{t,t_o}P$ will end up after a finite time at the boundary $\partial\Gamma = \partial\Omega\times\mathbb{R}^k$. How can $T_{t,t_o}P$ be globally defined?
Suppose t_1, t_o to be fixed times, $t_1 \geq t_o$. We are going to give a precise meaning to $T_{t_1,t_o}P$ for all starting points P. Denote by Γ^- the set of all points $P_o \in \Gamma$ for which $T_{t,t_o}P_o$ ends up at $\partial\Gamma$ at some time $t_-(P_o) \in (t_o,t_1)$:

$$T_{t_-(P_o),t_o}P_o \in \partial\Gamma.$$

Further, define by $\partial\Gamma^-$ the set of positions on $\partial\Gamma$ with V_x pointing outward of Γ (V_x are the first l components of V), and by $\partial\Gamma^+$ those with V_x pointing inward:

$$\partial\Gamma^{\pm} = \left\{P \in \partial\Gamma : \langle V_x(P), n(P)\rangle \gtrless 0\right\},$$

where $n(P)$ is the inner normal on $\partial\Omega$ at P_x (P_x are the position coordinates of P). For simplicity, we assume $\partial\Gamma^{\pm}$ to be time independent.

We define T_{t_1,t_o} as follows:

- On the complement $(\Gamma^-)^c$ of Γ^-, $T_{t_1,t_o}P$ is defined as before.
- On Γ^-, we define

$$T^- : \Gamma^- \longrightarrow (t_o,t_1] \times \partial\Gamma^-$$

by

$$T^-P_o := (t_-(P_o),\ T_{t_-(P_o),t_o}P_o).$$

If we assume

(A) Trajectories starting from the wall do not return to the wall any more, i.e.

$$P_o \in \partial\Gamma^+ \longrightarrow T_{t,s}\,P_o \notin \partial\Gamma^- \ \forall\ t \geq s,$$

then we may define T^+ on $(t_o,t_1] \times \partial\Gamma^+$ by

$$T^+\,(t,P) = T_{t_1,t}\,P,$$

so that we have

$$\Gamma^- \xrightarrow{T^-} (t_o,t_1] \times \partial\Gamma^- \quad (*) \quad (t_o,t_1] \times \partial\Gamma^+ \xrightarrow{T^+} \Gamma.$$

The gap at (*) has to be closed by a reflection law which is a set

$$\left\{P^{ref}_{t,a,v} \mid t \geq t_o, (a,v) \in \partial\Gamma^-\right\}$$

of probability measures on $[t,\infty] \times \partial\Gamma^-$ with certain properties (see Babovsky[3]).

A procedure like this has certain implications:

- if $P^{ref}_{t,a,v}$ are δ-measures (i.e. the reflection law is deterministic), then T_{t_1,t_o} may again be defined as a mapping from Γ to Γ, but it need not be a bijection.
- in the other case (i.e. if the reflection law is stochastic), then T_{t_1,t_o} as a mapping : $\Gamma \longrightarrow \Gamma$ makes no sense, since uniqueness is lost a the boundary. However, at least under certain restrictions on the reflection law, we may give sense to μ_t by defining

$$\mu^- := \mu_{t_o} \circ (T^-)^{-1}$$

$$\mu^+ := \int P^{ref}_{t,a,v} \;(.)\; d\mu^-(t,a,v),$$

$$\mu_{t_1} := \mu^+ \circ (T^+)^{-1}.$$

In the case of all distributions being absolutely continuous this is equivalent to solving the initial boundary value problem

$$\frac{\partial}{\partial t} f + \text{div}\,(f \cdot V) = 0$$

(IC) $f(t_o) = f_{t_o}$

(BC) $f_+(t,a,v) = \int p_{t',a',v'}(t,a,v)\; f_-(t',a',v')\; dt'd\omega(a')d^3v'$,

where $d\omega$ is the surface measure on $\partial\Omega$, and $f_\pm$ and $p_{t,a,v}$

are the densities of $\mu^{\pm}$ and $P_{t,a,v}$. Further, it is an easy matter to show that $f_{\pm}(a,v) = |\langle n(a),v\rangle| \cdot f(a,v)$.

- more complicated becomes the situation if we reject (A): Particles coming from the boundary may return to the boundary again. In this case, the gap at (*) has to be closed by chains of the form

$$\left(\xrightarrow{\text{pref}} (t_o,t_1] \times \partial\Gamma^+ \xrightarrow{T_{t,s}} (t_o,t_1] \times \partial\Gamma^- \right)^l \xrightarrow{\text{pref}}$$

l+1 representing the total number of contacts with the wall. Similarly as shown above, we may for each chain define a distribution $\mu_{t_1}^{(l)}$ (representing the particles with exactly l collision with the wall during the time intervall $(t_o,t_1]$) so that

$$\mu_{t_1} = \sum_{l=o}^{\infty} \mu_{t_1}^{(l)}$$

However, one should be aware of the facts that mass conservation and uniqueness are not guaranteed and that (BC) has to be interpreted in a weaker sense. A program like this has been carried out by Babovsky[3] for the simplest case: a Knudsen flow.

2.4 The velocity field depending on the distribution

In section 2.2. we derived the continuity equation for a particle flow in a given (exterior) field, for example an electro-magnetic field. The Vlasov equation describes exactly the same situation in the case that the field (or part of the field) is generated by the particle distribution itself.

Here, we discuss only one special case: we suppose the differential equation for $P = (x,v)$ to read

$$\text{(DE)}\qquad \begin{aligned} \dot{x} &= v \\ \dot{v} &= \gamma E \end{aligned}$$

(or equivalently: $V(t,P,\mu) = (v,\gamma E(P,\mu))$, where E depends on μ as follows:

$$E_t(x) = -\int \frac{x-y}{\|x-y\|^k} (d\mu_t(y,v) - n(y)dy) =: \nabla_x \Phi(x,\mu),$$

(k=1 is the dimension of the position space).

In the case of plasma physics, E_t is an electric field created by the ion distribution $d\mu_t$ and a given ion background $n(y)dy$; in stellar dynamics, E_t is the gravitational field. (γ is positive for repulsive and negative for attracting forces.)

As described by the transport theorem, the evolution equation for the densities $f_t = \frac{d\mu_t}{d\lambda}$ is

$$\text{(CE)}\qquad \frac{\partial}{\partial t} f + v\cdot\nabla_x f + \gamma E[f]\cdot\nabla_v f = 0 .$$

Combined with the equation for E, this is the Vlasov equation.

The most interesting questions are now:

- Is again the continuity equation (CE) a consequence of the differential equations (DE)?
- Is a continuous system derivable from an N-particle system, as N approaches infinity?
- Does there exist a unique solution of (CE)?

Things become quite simple if the singular term in the definition of E:

$$G = \frac{x-y}{\|x-y\|^k}$$

is replaced with some mollified kernel $G^\delta \in C^b(\mathbb{R}^k)$ where $G^\delta \longrightarrow G$ for $\delta \longrightarrow 0$. Since G^δ is globally Lipschitz continuous, the same is true for the mapping

$$P \longrightarrow V^\delta[\mu_t]$$

(V^δ being the mollified velocity field). Denote by C_M the set of all flows $\mu.$ of probability measures which are weakly continuous in t. Then for all $\mu. \in C_M$, $V^\delta[\mu.](P)$ is coninuous. Therefore, the initial value problem

$$\dot{P} = V^\delta[\mu_t](P), \quad P(s) = Q$$

defines a flow $T^\delta_{t,s}[\mu.]\,Q$ for all t,s, and the mollified Vlasov equation may be defined as a generalized continuity equation for measures:

$$\mu_t = \mu_o \circ T^\delta_{o,t}[\mu.].$$

As before we obtain: If μ_o is absolutely continuous then also μ_t, and $f_t = \frac{d\mu_t}{d\lambda}$ is a solution of

$$(CE)^\delta \qquad \frac{\partial f_t}{\partial t} + \mathrm{div}_p(f_t V^\delta) = \frac{\partial f_t}{\partial t} + v\cdot\nabla_x f_t + \gamma\cdot E_t^\delta\cdot\nabla_v f_t = 0$$

Existence and uniqueness for $(CE)^\delta$ follow essentially by application of some Banach fixed point argument. Furthermore, in the case of $\Omega = \mathbb{R}^k$ or some deterministic reflection law,

$$\mu_o^N \longrightarrow \mu_o$$

implies

$$\mu_t^N \longrightarrow \mu_t$$

as N tends to infinity. Thus, continuous systems are derivable from N-particle systems.For a more detailed discussion of these problems see Neunzert[4].

Let us have a short look at stochastic reflection laws: Starting from an N-particle distribution

$$\mu_o^N = \frac{1}{N} \Sigma\ \delta(P-P_j(0)),$$

the solution keeps for some time the form

$$\frac{1}{N} \Sigma\ \delta(P-P_j(t)).$$

However, at the first contact with the wall, the corresponding δ-distribution is smeared out. From now on, $E[\mu_t]$ is no longer a field generated by N particles. Thus the correct phase space in this case is not Γ but Γ^N. Until now (as far as we know) there has not been obtained any convergence result for this problem for $N \longrightarrow \infty$.

Here, we do not discuss the limit $\delta \longrightarrow 0$. Two possible ways are

- Use compactness arguments to prove weak convergence. In this case uniqueness is lost.
- Show convergence (for example of $T_{t,o}^{\delta}\ [\mu.^{\delta}]$) in the limit $\delta \longrightarrow 0$. Then one may obtain existence and uniqueness even for classical solutions. This is possible, but not in the case k = 3.

For more details we refer to several papers[4,5,6,7].

3. Stationary solutions of the Vlasov equation

3.1 Construction of explicit solutions

The special form of the Vlasov equation admits in simple cases the explicit construction of stationary and of time-periodic solutions.

At first, let's consider stationary non-homogeneous solutions in one dimension. The Vlasov equation for this case reads

$$v \cdot f_x - \Phi_x \cdot f_v = 0.$$

This equation is satisfied if we choose the ansatz

$$f(x,v) = F(v^2 + 2\Phi(x)).$$

For the potential, we have Poisson's equation

$$\Phi''(x) = \int_{-\infty}^{\infty} F(v^2 + 2\Phi(x))\, dv.$$

Given F, this is a 2nd order nonlinear differential equation for Φ. Or, given Φ, this is an integral equation for F. Let's be more careful concerning this last point (see Neunzert[8]). It is even easier to consider two kinds of particles. In normalizing m_i with respect to $\pm q_i$ we get

$$v \cdot \frac{\partial f_-}{\partial x} + E(x) \cdot \frac{\partial f_-}{\partial v} = 0$$

$$v \cdot \frac{\partial f_+}{\partial x} - E(x) \cdot \frac{\partial f_+}{\partial v} = 0$$

$$E'(x) = c \int f_-(x,v)dv - c_+ \int f_+dv = \rho(x).$$

Then

$$f_-(x,v) = F_-(v^2-2\Phi(x))$$

and

$$f_+(x,v) = F_+(v^2+2\Phi(x)).$$

Consider the interval (x_o,x_1) and an arbitrary $\xi \in (x_o,x_1)$. If $f_-(\xi,v)$ is given then $f_-(x,v)$ is given in $[\xi,x_1)$. Is $f_+(\xi,v)$ known then $f_+(x,v)$ is known for $|v| \geq \sqrt{2(\Phi(\xi) - \Phi(x))}$. If both distributions are given for $x = \xi$, then we know everything about ρ besides

$$\int_{|v| \leq \sqrt{2(\Phi(\xi) - \Phi(x))}} f_+(x,v)dv.$$

If in addition Φ is given, then $\Phi'' = \rho$ becomes

$$\int_{|v| \leq \sqrt{2(\Phi(\xi) - \Phi(x))}} F_+(v^2 + 2\Phi(x))dv = h(x) = \text{known function}.$$

Substituting $v^2 + 2\Phi(x) = t$ one gets

$$\int_{2\Phi(x)}^{2\Phi(\xi)} F_+(t) \frac{dt}{\sqrt{t-2\Phi(x)}} = h(x).$$

Since Φ is strictly monotone in $[\xi,x_1]$, this is Abel's integral equation which can be solved:

$$F_+(z) = \frac{1}{\pi}\frac{d}{dz} \int_z^{2\Phi(\xi)} \frac{h(\psi(t))}{\sqrt{z-t}}\, dt, \quad 2\Phi(x) \leq z \leq 2\Phi(\xi)$$

where ψ is the inverse function to $2\Phi(x)$.

Therefore: $f_\pm(\xi,v)$ and $\Phi(x)$ given in $[\xi,x_1]$ determine $f_\pm(x,v)$ in $[\xi,x_1]$. But $f_\pm(x,v)$ together with $\Phi(x)$ in $[x_1,x_2]$ determine $f_\pm(x,v)$ in $[x_1,x_2]$ etc. Finally: $f_\pm(\xi,v)$ and Φ given, there exists a unique solution $f_\pm$ which - unfortunately - is not necessarily positive. But it can be constructed explicitely (BGK-modes). The solutions obtained (it's a big variety) have been observed in experiments.

Let us consider the other idea: Given F we get an ODE for Φ in a more general geometric situation (Batt el al[9]). (Here, we consider only one particle species.) In the spherically symmetric case, f depends only on $\|x\| = r$, $\|v\| = w$ and the angle between x and v, for example on

$$\varphi := \|x\|^2 \cdot \|v\|^2 - \langle x,v\rangle^2.$$

f is a stationary solution if

$$f(x,v) = \tilde{f}(r,w,\varphi) = F(w^2 + 2\Phi(r) + \frac{\varphi}{r^2}, \varphi).$$

Now,

$$\rho(x) = \tilde{\rho}(r) = \frac{\pi}{r^2}\int_{-\infty}^{\infty}\int_0^{\infty} F(w^2+2\Phi(r) + \frac{\varphi}{r^2}, \varphi)\, d\varphi\, dw =: h_\Phi(r)$$

and Poisson's equation reads

$$\frac{1}{r^2}(r^2\Phi')' = h_\Phi(r).$$

Batt et al. discuss existence and uniqueness of this second order ODE with the boundary condition

$$\lim_{r \to 0} \Phi(r) = \alpha.$$

The problem of positivity is trivial in this case, but one gets only existence results, no method of construction of solutions.

3.2 Stability of stationary solutions

We have seen that stationary inhomogeneous solutions can be constructed mathematically. But do they also exist physically, i.e. can they be observed? This leads to the question whether stationary solutions are stable. There are not many results concerning stability. Here, we will shortly review two of them.

Before, we have to say a few words about what kind of stability might be proven:

Linearized stability concerns solutions of an equation which is linearized about a given stationary solution.

Nonlinear stability concerns solutions of the full equation. There are cases in which one may conclude nonlinear stability from linear stability, but there are also cases one can not. In general we mean by stability (nonlinear or linear) of a stationary solution μ_0 of an equation the following:

For all $\varepsilon > 0$, there is a $\delta(\varepsilon) > 0$ such that

$$\| \tilde{\mu}(0) - \mu_0 \| < \delta$$

implies

$\|\tilde{\mu}(t) - \mu_o\| < \varepsilon$ for all t.

Here, $\tilde{\mu}$ is any other (in general instationary) solution of the equation.

We are now going to sketch a result by Marchioro and Pulvirenti[10]. They study a continuum of electrons in a k-dimensional torus $\Omega = T^k$ with a uniform positive background (such that the system is neutral). Only stationary solutions of the form

$$f_o(x,v) = \Phi(\|v\|) : \mathbb{R}_+ \longrightarrow \mathbb{R}_+$$

are considered, where Φ is bounded and nonincreasing. Further it is assumed that

$$\begin{cases} \int v^2 \Phi dv < \infty, & \text{if } k = 1,2 \\ \Phi \text{ has compact support, } & \text{if } k = 3. \end{cases}$$

Define for any $M > 0$

$$J_M(f_o) := \left\{ f \,\middle|\, \int |f-f_o| \cdot \|v\|^2 dv \leq M \right\}.$$

Then the result is: If the initial condition $f^{(o)}$ lies in $L_\infty \cap J_m(f_o)$ for some M, then for every $\varepsilon > 0$ there exists a δ such that

$$\|f_o - f(0)\|_1 < \delta$$

implies

$$\|f_o - f(t)\|_1 < \varepsilon.$$

The proof uses conservation of energy:

$$E_{tot}[f(t)] = \frac{1}{2}\,\|E_t\|_2^2 + \frac{1}{2}\int\|v\|^2 f(t,x,v)\,dv\,dx$$
$$= U[f(t)] + E_{kin}[f(t)].$$

(U is the internal and E_{kin} the kinetic energy).
Now define

$$I(f_o) := \Big\{f\,\Big|\text{ for all } \alpha \in \mathbb{R}_+,\ \lambda(\{(x,v)\,|\,f(x,v) \geq \alpha\}) = \lambda(\{(x,v)\,|\,f_o(x,v)\geq\alpha\})\Big\}.$$

Liouville's theorem tells us that $f_t \in I(f_o)$ if f(0) is. For arbitrary f(0) with $\|f(0) - f_o\|_1 < \delta$ there exists always an $f^*(0)$ of the form $f^*(0,x,v) = \psi(\|v\|)$, ψ nondecreasing, which is in the same class, i.e. $f(0) \in I(f^*(0))$, and also

$$\|f^*(0) - f_o\|_1 < \delta.$$

Next it is shown that for $f(0) \in I(f_o)$ we get

$$E_{kin}[f(t)] - E_{kin}[f_o] \leq g(\|f(0) - f_o\|_1)$$

with $g(z) \downarrow 0$ for $z \downarrow 0$ monotonically nonincreasing.
A further lemma gives (for k = 2,3) for $f \in I(f_o) \cap L_\infty$

$$\|f - f_o\|_1^2 \leq c\cdot(E_{kin}[f] - E_{kin}[f_o])$$

(this is a nice elementary integration lemma!).
The rest is simple:

$$\|f(t) - f_o\|_1 \leq \|f(t) - f^*(0)\|_1 + \|f^*(0) - f_o\|_1$$
$$\leq c\cdot(E_{kin}[f(t)] - E_{kin}[f^*(0)])^{1/2} + \|f^*(0) - f_o\|_1$$
$$\leq c\cdot g(\|f(0) - f^*(0)\|)^{1/2} + \|f^*(0) - f_o\|_1$$

$\leq c\cdot g(2\delta)^{1/2} + \delta$, if $\|f(0) - f_o\|_1 < \delta$

which completes the proof.

Not very much is known about the stability of BGK-modes-probably they are in general nonlinearly unstable. The second paper we are going to review is the dissertation of Pfreundt[11] in which the author intends to study also inhomogeneous stationary solutions. However, because of the complexity he restricts to model equations.
Starting point is the ansatz

$$f(t,x,v) = \sum_{\substack{n\in Z \\ m\in N}} c_{nm}(t)\, e^{2\pi inx}\, H_m(v) e^{-v^2}$$

where H_m are the Hermite polynomials.
Using well-known properties of the Hermite polynomials, especially the orthogonality relations, one obtains from the Vlasov-Poisson system a coupled system of ordinary differential equations for c_{nm}:

$$\dot{c}_{nm} = \sum a^{nm}_{klrs}\, c_{kl}\, c_{rs}.$$

From these equations one may easily obtain some iteration procedure to construct stationary solutions.
Now the system is truncated: $|n| \leq 2$, $m \leq 2$. The results are as follows:

<u>Linear stability</u>: The eigenvalues of the corresponding functional matrix are analyzed, and it is shown that

- for the homogeneous Maxwell distribution all eigenvalues are on the imaginary axis. Thus, a stability result has to come from nonlinear theory;
- bounded away from the Maxwellian, all stationary solutions are unstable. This is due to the fact that $\bar{\lambda}$ and

$-\lambda$ are eigenvalues if λ is. So if not all eigenvalues are on the imaginary axis, there are eigenvalues with positive and such with negative real part.

To see whether Maxwellians are stable,

Nonlinear stability has to be studied: It is shown that all stationary solutions are at least asymptotically unstable. Further, numerical experiments seem to indicate that also the Maxwellians are nonlinearly unstable in the sense defined above.

What may be concluded from these results for the truncated system (which is no Hamiltonian)? Certainly we may not transfer them directly to the full Vlasov equation. For this, the model is too rough. However, they indicate what might be expected for the Vlasov equation:

First of all, what we have to expect, is a very complicated dynamics. Second, it seems that at least inhomogeneous solutions should be unstable, perhaps except those which are slight perturbations of homogeneous solutions.

But let's repeat it: The results are only an indicator, not a proof.

4. A Vlasov Equation in Nuclear Physics

Nowadays, the Vlasov equation also appears in a completely different context: in nuclear physics. We are now going to introduce the "nuclear" Vlasov equation, to show the properties common with the classical one and also the differences, and to try to carry over the "know how" for the classical case (which has been collected through many years) to the nuclear case.

4.1 A "Vlasov-like" equation

The foundations of what follows now may be taken from Bertsch[12]. The evolution of the wave function $\psi(t,x_1,\ldots,x_N)$ of an N-particle system is given by

$$i\,\frac{\partial\psi}{\partial t} = \Big(-\frac{1}{2m}\sum_{j=1}^{N}\Delta x_j + \sum_{1\le i<j\le N}\Phi(x_i-x_j)\Big)\cdot\psi.$$

For the density matrix,

$$\tilde{\rho}(t;x_1,\ldots,x_N;\,x_1',\ldots,x_N')=\bar{\psi}(t;x_1,\ldots,x_N)\cdot\psi(t;x_1',\ldots,x_N')$$

one easily derives

$$-i\,\frac{\partial\tilde{\rho}}{\partial t} = \Big(-\frac{1}{2m}\sum_{j=1}^{N}(\Delta_{x_j}-\Delta_{x_j'}) + \sum_{i<j}(\Phi(x_i-x_j)-\Phi(x_i'-x_j'))\Big)\cdot\tilde{\rho}.$$

For the integrated densities

$$\rho^{(1)}(t,x,y) := N\cdot\int\tilde{\rho}(t;x,z_2,\ldots,z_N;y,z_2,\ldots,z_N)\,dz_2\ldots dz_N$$

and

$$\rho^{(2)}(t;x_1,x_2,y_1,y_2) := N\cdot(N-1)\int\tilde{\rho}(t;x_1,x_2,z_3,\ldots,z_N;\,y_1,y_2,z_3,\ldots,z_N)\,dz_3\ldots sz_N$$

the following relation holds:

$$-i\,\frac{\partial\rho^{(1)}}{\partial t} = -\frac{1}{2m}(\Delta_x-\Delta_y)\rho^{(1)} + \int[\Phi(x-z)-\Phi(y-z)]\cdot\rho^{(2)}(t;x,z,y,z)\,dz.$$

Introducing the Hartree-Fock-approximation:

$$\rho^{(2)}(t;x,u,y,v) = \rho^{(1)}(t,x,y)\cdot\rho^{(1)}(t,u,v) - \rho^{(1)}(t,x,v) \cdot \rho^{(1)}(t,u,y)$$

one ends up after the substitutions

$$\xi := \frac{x+y}{2}, \quad \eta := \frac{x-y}{2}, \; \rho^*(\xi,\eta) := \rho(x,y)$$

with

$$-i\,\frac{\partial\rho^*}{\partial t} = -\frac{1}{2m}\sum_{l=1}^{k}\frac{\partial^2}{\partial\xi_l\partial\eta_l}\rho^* + V^{(1)}[\rho^*]\cdot\rho^* - V^{(2)}[\rho^*,\rho^*],$$

where

$$V^{(1)}[\rho^*](\xi,\eta) = \int[\Phi(x-z) - \Phi(y-z)]\cdot\rho^*(1,z,0)\,dz$$

and

$$V^{(2)}[\rho^*,\rho^*](\xi,\eta) = \int[\Phi(x-z) - \Phi(y-z)]\cdot\rho^*(\tfrac{1}{2}(x+z),\tfrac{1}{2}(x-z))\cdot \rho^*(\tfrac{1}{2}(z+y),\, \tfrac{1}{2}(z-y))\,dz.$$

Take the Fourier transform $f(t,\xi,\zeta)$ with respect to η ("Wigner transform") to obtain

$$\frac{\partial f}{\partial t} + \frac{1}{2m}\,\zeta\cdot\nabla_\xi f = i\int(V^{(1)*}[\rho^*]\cdot\rho^* - V^{(2)*}[\rho^*,\rho^*])\cdot e^{i\langle\eta,\zeta\rangle}d\eta$$

With $U[\rho](x) = \int\Phi(x-z)\rho(t,z,z)dz$ we have

$$V^{(1)}[\rho](x,y) = U[\rho](x) - U[\rho](y);$$

with

$$U[\rho]_{exch}(t,x,y) = \int \Phi(x-z)\rho(t,x,z)\rho(t,x,y)dz$$

one gets

$$V^{(2)}[\rho,\rho](t,x,y) = U[\rho]_{exch}(t,x,y) - U[\rho]_{exch}(t,y,x).$$

Further, a mostly used approximation substitutes $U[\rho]_{exch}$ (t,x,y) by

$$\rho(t,x,y)\cdot\int\Phi(x-z)\rho_{FG}(z,z)dz$$

where ρ_{FG} is the density matrix of the infinite Fermi gas, depending again in an explizit way on $\rho(t,z,z)$.
Collecting everything, denoting

$$U_{tot}[\rho](t,x) = \int\Phi(x-z)[\rho(t,z,z) + \rho_{FG}(z,z)]dz$$

we end up with the equation

$$\frac{\partial f}{\partial t} + \frac{1}{2m}\,\xi\cdot\nabla_{\xi} f - P_U[f] = 0$$

which resembles the Vlasov equation if $P_U[f]$ can be given a similar shape as the term of the Vlasov equation containing the potential.
$P_U[f]$ is defined by

$$P_U[f](t,\xi) = \frac{1}{(2\pi)^k}\iint e^{i(v-w)z}\sigma_{\xi}(w,z)f(t,z,w)dzdw$$

with the "symbol"

$$\sigma_x(w,z) := i[U_{tot}[f](x+\tfrac{z}{2}) - U_{tot}[f](x-\tfrac{z}{2})]$$

(not depending on w).

4.2 The nuclear Vlasov equation

We obtain the nuclear Vlasov equation if we assume an asymptotic expansion for $\sigma_x(z)$:

$$\sigma_x(z) = \sum_{j=0}^{\infty} \sigma_x^{(j)}(z),$$

where $\sigma_x^{(j)}$ is homogeneous in z of the order (1-j). From the definition of σ_x we get

$$\sigma_x^{(0)}(z) = iz\cdot\nabla_x U_{tot},$$

which we call the principal symbol. If such an expansion is justified, then P_U is a pseudo-differential operator of order 1, and as a $0^{\underline{th}}$ order approximation we may replace P_U by

$$P_U^{(0)}[f](z) = \frac{1}{(2\pi)^k}\iint e^{i<v-w,z>}\sigma_x^{(0)}(z)\cdot f(t,z,w)\,dzdw$$

$$= \nabla_x U_{tot}\cdot\nabla_v f;$$

the result is the nuclear Vlasov equation

$$\frac{\partial f}{\partial t} + \frac{1}{2m}\, v\cdot\nabla_x f - \nabla_x U_{tot}\cdot\nabla_x f = 0.$$

(Here, ξ and ζ have been replaced by x and v.)

There are certain models in which U_{tot} takes the form

$$U_{tot}[f] = c\cdot\int \Phi(x-y)\cdot[\int f(t,y,v)dv + \varsigma_{FG}(y,y)]dy$$

or corresponds to the "Bonche-Koonin-Negele" effective interaction:

$$U_{tot}[n](t,x) = \alpha n(t,x) + \beta n^2(t,x) + \gamma \int \Phi(x,y)n(t,y)dy.$$

4.3 Classical versus nuclear Vlasov equation

There are certain differences between the classical and the nuclear Vlasov equation which may not be overlooked:

- f is no density in the classical sense, so f need not be positive.
- The dependence of $n \longrightarrow U_{tot}[n]$ is nonlinear, there is no analogue of Poisson's equation for U_{tot}.
- There are "local" parts (for example in the Bonche-Koonin-Negele approach if α and β do not vanish).

What can be carried over? Until now, not very much has been done explicitly. We are shortly discussing several ideas. For a more detailed presentation see Neunzert[13].

a) Construction of stationary solutions:
In the classical case, the one-dimensional and the spherically symmetric case are well developed[8,9]. Inserting an ansatz as shown before, we arrive at relations

$$n(x) = \int_{-\infty}^{\infty} F(v^2 + 2U_{tot}(x))dv \quad \text{in the 1D case}$$

and

$$n(r) = \frac{\pi}{r^2} \iint F(w^2 + 2U_{tot}(r) + \frac{\Phi}{r^2}, \Phi)\, dw\, d\Phi$$

in the spherically symmetric case.

The problem is - of course - that U_{tot} depends nonlinearly on n. There are several proposals to overcome this difficulty[13].

b) Existence and uniqueness for solutions of the IVP:

In the classical case, there are certain existence and uniqueness theorems which are based on the Lipschitz coutinuity of γE and a condition like the following:

$$\text{(LC)} \qquad \| E[\mu] - E[\tilde{\mu}] \|_\infty \leq K \cdot d(\mu, \tilde{\mu}),$$

where d is some distance in the space of distributions. A convenient distance is the "bounded Lipschitz distance"

$$d(\mu,\tilde{\mu}) = \sup_{\Phi \in D} \left| \int \Phi(x)\, [d\mu - d\tilde{\mu}] \right|$$

with $D = \left\{ \Phi : \mathbb{R}^k \longrightarrow [0,1] \mid |\Phi(x) - \Phi(y)| \leq |x-y| \right\}$.

For this distance, the equation above implies

$$\text{(LC)'} \qquad \| E[\mu] - E[\tilde{\mu}] \|_\infty \leq K \cdot \| f - \tilde{f} \|_1$$

if f and $\tilde{f}$ are the densities of μ and $\tilde{\mu}$.

So for the nuclear equation linearity of $\nabla_x U_{tot}$ is not so important, but in general (LC) is not satisfied. One can see this for the Bonche-Koonin-Negele interaction where (LC)' cannot be satisfied because of the local parts. However, at least mollified equations can be solved.

c) Numerical schemes:
Finally, one should realize that there have been developed a lot of numerical schemes for the classical case (finite difference, Galerkin, Particle-in-cell methods, etc.[14,15]) which might be of interest for the nuclear equation.

Besides all these proposals, however, one has to face the problem of how all the approximations in the derivation, especially the truncation of σ_X involve the validity of the nuclear Vlasov equation.

We think there are a lot of open questions about the nuclear Vlasov equation, and a really big progress can be obtained if one understands how to apply the methods of the classical Vlasov equation.

References:

1. Billingsley, P., **Convergence of Probability Measures**, J. Wiley & Sons, New York, 1968.
2. Rademacher, H., Eineindeutige Abbildungen und Meßbarkeit, **Monatsh. Math. Phys.** 27, 183, 1916.
3. Babovsky, H., Initial and boundary value problems in kinetic theory. I. The Knudsen gas, **Transp. Theory Stat. Phys.** 13, 455, 1984.
4. Neunzert, H., Mathematical investigations on particle-in-cell methods, in **Fluid Dynamics Transactions** 9, Warzawa, 1978.
5. Illner, R. and Neunzert, H., An existence theorem for the unmodified Vlasov equation, **Math. Meth. Appl. Sci.** 1, 530, 1979.
6. Horst, E., Global strong solutions of Vlasov's equation - necessary and sufficient conditions of their existence, to appear in **Banach Center Publications**, Warsaw.
7. Horst, E. and Hunze R., Weak solutions of the initial value problem for the unmodified non-linear Vlasov equation, **Math. Meth. Appl. Sci.** 6, 262, 1984.
8. Neunzert, H., **Über ein Anfangswertproblem für die stationäre Boltzmann-Vlasov Gleichung**, Berichte der KFA Jülich Nr. 297, 1965.
9. Batt, J., Faltenmacher, W. and Horst, E., Stationary spherically symmetric models in stellar dynamics, **Arch. Rat. Mech. Anal.** 93, 159, 1986.
10. Marchioro, C. and Pulvirenti, M., A note on the non-linear stability of a spatially symmetric Vlasov-Poisson flow, **Math. Meth. Appl. Sci.** 8, 284, 1986.

11. Pfreundt, F.J., **Zur Stabilität von stationären Lösungen der Vlasov-Poisson-Gleichung**, Dissertation, Universität Kaiserslautern, 1986.

12. Bertsch, G.F., **Heavy ions dynamics at intermediate energy**, Les Holmes Lectures, 1977 (North Holland, 1978).

13. Neunzert, H., **The "nuclear" Vlasov equation - methods and results that can (not) be taken over from the "classical" case**, preprint, Universität Kaiserslautern, 1984.

14. **Methods in Computational Physics**, Vol. IX, Academic Press, New York, 1970.

15. Neunzert, H., Approximation methods for the nonmodified Vlasov-Poisson system, in Proceedings of **Mathematical Aspects of Fluid and Plasma Dynamics**, Triest, 1984.

EXISTENCE AND UNIQUENESS THEOREMS FOR THE BOLTZMANN EQUATION

A. Palczewski
Warsaw University, Warsaw, Poland

1. THE BOLTZMANN EQUATION

The aim of these lecture notes is the presentation of existence theorems for the Boltzmann equation. We do not attempt to give a complete survey of the present state of the theory as two such surveys has been written recently (Fiszdon, Lachowicz, Palczewski [12] and Greenberg, Polewczak, Zweifel [17]). We are rather going to present those results which, according to the present author's opinion, shown to be most fruitful in the further development of the whole theory.

Let us recall, to start with, the classical formulation of Boltzmann's equation concerning the evolution of the one particle distribution function $f=f(x,v,t)$ of a monoatomic dilute gas:

$$\partial f/\partial t + v\cdot \mathrm{grad}_x\, f = J(f,f), \tag{1.1}$$

where x,v are the position and velocity vectors and t is time. If A_x, A_v are measurable subsets of R^3 then:

$$\int_{A_v}\int_{A_x} f(x,v,t)\, dx dv$$

is interpreted as the average number of particles in A_x with velocities in A_v at time t.

$$J(f,f)(x,v,t) = \iiint k(|v-v_1|,\theta)\cdot\{f(x,v',t)\cdot f(x,v_1',t) - \\ - f(x,v,t)\cdot f(x,v_1,t)\}\, d\varepsilon d\theta dv_1 \tag{1.2}$$

is the collision integral, where $|v-v_1|^{-1}\cdot k(|v-v_1|,\theta)$ is the collision cross section, post collisional parameters are primed, $\chi = \pi - 2\theta \in [0,\pi]$ is the scattering angle of the binary collisions, $\varepsilon \in [0,2\pi]$ is the azimuthal angle of the plane in which the collisions take place.

The collision process depends strongly on the particle interaction potential $U(r)$ which, for the spherically symmetric particles considered depends only on their distance apart, r. For interparticle potentials of the form $U(r)=r^{-n}$ the collision kernel has the form:

$$k(|v-v_1|,\theta) = \mathrm{const}\cdot |v-v_1|^{1-4/n}\cdot \beta_n(\theta),$$

where $\beta_n(\theta)$ - the differential collision cross section. It can be seen that for n=4 the collision rate becomes independent of the relative velocity $|v-v_1|$ and then this interaction law corresponds to Maxwell's molecules. Particle interaction potentials with $n \le 4$ are called "soft" and for $n>4$ "hard" interaction potentials. The model of rigid spherical molecules, for which $k(|v-v_1|,\theta)=\text{const}\cdot|v-v_1|\cdot\sin\theta\cdot\cos\theta$, is included to "hard" interactions ($n\to+\infty$). For the rigid spheres model the collision operator splits as $J(f,f)= Q(f,f)-f\cdot P(f)$. This splitting does not hold for power interparticle potentials because $\beta_n(\theta)$ is not integrable over $[0,\pi/2]$, as β_n has a non-integrable singularity at $\theta=\pi/2$, which occurs for so-called grazing collisions. This mathematical difficulty does not occur for cut-off potentials. Most often the following cut-offs are used:

a) angular cut-off (Grad [15]),
b) radial cut-off (Cercignani [8]),
c) integral cut-off (Drange [9]).

This survey of mathematical problems of the Boltzmann equation is mainly devoted to the existence and properties of solutions of equation (1.1). In what follows we shall use several definitions of solutions of this equation. By classical solution of (1.1) we shall understand a function $f(x,v,t)$ which is continuously differentiable with respect to x,v and t variables and which fulfills the equation in the classical sense. We shall consider also equation (1.1) in Banach spaces of functions of x and v variables (or v variable only). Let B be one of such spaces. A distribution function $f(x,v,t)$ can be considered as a trajectory $f(t)$ in B and the term $v\cdot\text{grad}_x$ is just an unbounded operator in B. We shall call $f(t)$ a strong solution of (1.1) in B if $f(t)$ is a strongly differentiable trajectory in B and fulfills (1.1) in the norm of B.

We shall also use the notion of a mild solution. To define it let us write the transformation

$$T_\sigma(x,v,t) = (x+\sigma v, v, t+\sigma).$$

The function $f \cdot T_\sigma$ is called a mild solution of (1.1) if it is differentiable with respect to σ and satisfies the following equation:

$$d(f \cdot T_\sigma)/d\sigma = J(f,f) \cdot T_\sigma.$$

We begin the analysis of the Boltzmann equation recalling several properties of the collision integral J(f,f). First, let us recall that functions:

$$\begin{aligned} \psi_0 &= 1, \\ \psi_i &= v_i, \ i=1,2,3 \qquad (1.3) \\ \psi_4 &= v^2, \end{aligned}$$

called collision invariants, fulfill the equation

$$\int \psi_\alpha J(f,f)\, dv = 0, \ \alpha = 1,2,3,4, \qquad (1.4)$$

which corresponds to the preservation of mass, momentum and energy in a collision process.

On the basis of (1.4) it is easy to prove that the equation:

$$J(\omega,\omega) = 0 \qquad (1.5)$$

possesses the only solution of the form:

$$\omega = m/(2\pi T)^{-3/2} \exp(-|v-u|^2/2T). \qquad (1.6)$$

The function ω is called a Maxwell distribution function. When m, u and T are functions of x and t, ω is a local Maxwellian, otherwise it is called a global Maxwellian.

It is easy to check that a global Maxwellian is a stationary solution of the Boltzmann equation (1.1). Taking this into account we can look for solutions of (1.1) which are small perturbations of a global Maxwellian. From technical reasons we assume the solution in the form:

$$f = \omega + \omega^{1/2}F \qquad (1.7)$$

Insert this into (1.1) and neglect terms of second order in F to obtain the linearized Boltzmann equation:

$$\partial F/\partial t + v \cdot \mathrm{grad}_x F = LF, \qquad (1.8)$$

where

$$LF = \omega^{-1/2}J(\omega,\omega^{1/2}F). \quad (1.9)$$

We are now going to describe the properties of the linearized collision operator L. The existence of collision invariants ψ_α leads to the degeneracy of the null space of L. By simple calculations we can check that the functions:

$$\varphi_\alpha = \psi_\alpha\, \omega^{1/2}, \quad \alpha=0,1,2,3,4, \quad (1.10)$$

fulfill the equation:

$$L\varphi_\alpha = 0. \quad (1.11)$$

Hence the null space of L is fivefold degenerate.

For cut-off potentials the operator L, which acts only on the v-dependence of F, can be decomposed as follows:

$$L = -\nu + K, \quad (1.12)$$

where ν is the operator of multiplication by $\nu(v)$ and K is an integral operator. For soft potentials the function $\nu(v)$ satisfies:

$$0 < \nu(v) < \nu_0 < +\infty. \quad (1.13)$$

For hard potentials we have:

$$c_1(1 + |v|)^\gamma < \nu(v) < c_u(1 + |v|)^\gamma, \quad (1.14)$$

where $\gamma = 1-4/n$.

Usually L is considered in $L^2(R^3)$, where it is a nonpositive selfadjoint operator. Additionally, for hard potentials we have:

$$(f,Lf) < -\mu(f,f) \quad (1.15)$$

provided $(f,\varphi_\alpha) = 0$, $\alpha=0,\ldots 4$.

Using properties of L it is easy to prove that the operator

$$Bf = -v\cdot \mathrm{grad}_x f + Lf \quad (1.16)$$

is dissipative in $L^2(\Omega_x \times R_v^3)$, where $\Omega_x = R^3$ or is a bounded rectangular domain with periodic boundary conditions. The dissipativeness of B implies that B generates a strongly continuous contraction semigroup. This is a very useful property if we want to solve the nonlinear Boltzmann equation, however, L^2 - space is not appropriate to deal with the nonlinear problem. Hence a number of attempts has been made to extend this result on Banach spaces which are Banach algebras and where the nonlinear equation is easy to deal with. In connection with this we should mention the works of Ukai [32], Imai and Nishida [26], Shizuta [30], Caflisch [5] and Palczewski [28].

2.THE SPATIALLY HOMOGENEOUS BOLTZMANN EQUATION

In the case of spatially uniform problems the distribution function is independent of the space variables and the Boltzmann equation is then greatly simplified:

$$\partial f/\partial t = J(f,f),$$
$$f(v,0) = f_0(v), \tag{2.1}$$

where f_0 is the initial distribution function.

In the discussion of the existence problem for this equation the following spaces will be used: $B_q{}^r$ $(1\le q\le +\infty)$ is the space of measurable real functions on R^3 with the norm:

$$N_q{}^r\{f\} = (\int(1+|v|^2)^{r/2}|f(v)|^q dv)^{1/q} \text{ for } 1\le q<+\infty$$

and

$$N_\infty{}^r\{f\} = \operatorname*{ess\,sup}_v (1+|v|^2)^{r/2}|f(v)|.$$

$B_q{}^0$ and $N_q{}^0$ we denote simply L^q and $\|\cdot\|_q$ respectively. $C^{0,r}$ is the space of continuous real functions on R^3 with the norm:

$$N^r\{f\} = \sup_v (1+|v|^2)^{r/2}|f(v)|.$$

The first successful attempt to prove the existence and uniqueness for this equation was made by Carleman [6,7] for the case of rigid spherical molecules and an axially symmetric distribution function in velocity space. His result was improved by Maslova and Tchubenko [20,21,22] for the case of cut-off hard potential and an axially non-symmetric distribution function. The same result was obtained by Gluck [13].

<u>Theorem 1</u>

Let $n>4$ and

$$c_l < B_n(\theta)/(\sin\theta\cos\theta) < c_u \tag{2.2}$$

(a cut-off hard potentials or rigid spherical molecules).

Let $c_0 = \iint \beta_n(\theta)\, d\theta d\varepsilon$. If f_0 is non-negative and

$$f_0 \in C^{0,r} \text{ for } r < \max\{6,\ 3-4/n+4\pi c_u/c_0\} \qquad (2.3)$$

then there exists (for all $t>0$) a unique non-negative solution $f(t) \in C^{0,r}$ of (2.1) such that

$$\sup_{t>0} N^r\{f(t)\} < \text{const} < +\infty. \qquad (2.4)$$

■

To prove Theorem 1 we can write (1.1) in the form:

$$\partial f/\partial t = Q(f,f) - f \cdot P(f).$$

Taking $f = F \cdot \exp(\gamma t)$ we have

$$\partial F/\partial t + P_\gamma(F) \cdot F = Q_\gamma(F,F),$$

where

$$P_\gamma(F) = \gamma + \exp(\gamma t)P(F),\quad Q_\gamma(F,F) = \exp(\gamma t)Q(F,F).$$

Let now

$$V_\gamma(t_0\,;G,F) = G(v,t_0)\exp\{-\int_{t_0}^{t} P_\gamma(F)ds\} + \int_{t_0}^{t} Q_\gamma(F,F)\exp\{-\int_{\sigma}^{t} P_\gamma(F)ds\}d\sigma \qquad (2.5)$$

Then

$$F = V_\gamma(0;f_0,F) \qquad (2.6)$$

is the integral form of (2.1).

The following two lemmas establish properties of V_γ.

<u>Lemma 1</u>

If $t_0 \in [0,\ln 2/\gamma]$, $N^r\{G\}<+\infty$, $N^r\{F\}<+\infty$, where $r>5$ and $F>0$ for $t \in [0,\ln 2/\gamma]$ and $\delta>0$, we have:

$$N^r\{V_\gamma(t_0;G,F)\} <$$
$$< \max\{N^r\{G\},\ \sup_{t\in[0,\ln 2/\gamma]} N^r\{F\}(g+c_1N^r\{F\}/\gamma)\}, \qquad (2.7)$$

where $g = 4\pi c_u(1+\delta)/(r-2)c_0$ and c_1 is some constant.

■

<u>Lemma 2</u>

If F_1, G_1, F_2, $G_2 > 0$ and $N^r\{F_i\} < c < +\infty$, $N^r\{G_i\} < +\infty$ $(i=1,2)$, where $r>6-4/n$, $t \in [0,\ln 2/\gamma]$ then for $\delta > 0$ we have:

$$N^{r-1+4/n}\{V_\gamma(t_0;G_1,F_1)-V_\gamma(t_0;G_2,F_2)\} <$$

$$<\max\{N^r\{G_1-G_2\}, \sup_{t\varepsilon[0,\ln 2/\gamma]} N^{r-1+4/n}\{F_1-F_2\}(p+c_2\cdot c/\gamma)\}, \quad (2.8)$$

where $p = 4\pi c_u(1+\delta)/(r-3+4/n)c_0$, c_2 - some constant.

■

Proof of Theorem 1

Let $F_0 = f_0$ and $F_n = V_\tau(0;f_0,F_{n-1})$. We take $\delta>0$ such that $p<1$.Then $g<1$ as well. Let $\gamma_0=\max\{c_1,c_2\}\cdot N^r\{f_0\}\cdot(1-p)^{-1}$. Then for $\gamma>\gamma_0$ and $t\varepsilon[0,t_1] \subset [0,\ln 2/\gamma)$ we have $N^r\{F_n\}<N^r\{f_0\}$ and $N^r\{F_{n+1}-F_n\}<\alpha N^r\{F_n-F_{n-1}\}$ where $\alpha=p+c_2\cdot N^r\{f_0\}\gamma^{-1}<1$ for $r>3-4/n+4\pi c_u/c_0$. In this way we obtain a solution f of (2.1) as the limit of the sequence $F_n\cdot\exp(\gamma t)$ for $t \varepsilon [0,t_1]$ $[0,\ln 2/\gamma)$. Moreover $N^r\{f(t)\} < \exp(\gamma t)N^r\{f_0\}$.

Next, we show that $N^r\{f(t)\} < \text{const}$, where the constant depends only on $N^r\{f_0\}$. This enables us to extend a solution for an infinite time interval. Namely, in the interval $[t_1,2t_1]$ we solve the Boltzmann equation with the initial data $f(t_1)$ and by induction we have the solution for all $t>0$. Uniqueness of the solution follows from (2.8).

■

Another approach was made by Morgenstern [24,25], who considered (2.1) in L^1 and proved for cut-off Maxwell molecules that there exists a unique weak solution of (2.1) globally in time, in L^1, provided the initial data are non-negative and belong to L^1. Povzner [29] also investigated solutions in L^1 for the case of continuous $k(|v-v_1|)$ and proved that if $f_0 \varepsilon B_1^2$ then there exists a solution of (2.1), and for $f_0 \varepsilon B_1^4$ this solution is unique. An improvement of these results was obtained by Arkeryd [1].

First, let us assume:

$$k(|v-v_1|,\theta) < c(1+|v|^\lambda+|v_1|^\lambda) \text{ for } \lambda \varepsilon [0,2], \quad (2.9)$$

(this includes cut-off hard potentials and rigid spherical molecules). Then the result of Arkeryd can be stated as follows:

<u>Theorem 2</u>

If (2.9) is satisfied for some $\lambda\varepsilon[0,2)$ and $f_o>0$, $f_o\ln f_o\varepsilon L^1$, $f_o\varepsilon B_1^r$ for r=2 then there exists a non-negative weak (in L^1) solution f(t) of (2.1) for all t>0 such that:

$$f(t)\ \varepsilon\ B_1^r,\ t>0, \tag{2.10}$$

$$\int f(v,t)\,dv = \int f_o(v)\,dv, \qquad t>0, \tag{2.11a}$$

$$\int v\cdot f(v,t)\,dv = \int v\cdot f_o(v)\,dv, \qquad t>0, \tag{2.11b}$$

$$\int |v|^2\cdot f(v,t)\,dv \le \int |v|^2\cdot f_o(v)\,dv, \quad t>0. \tag{2.11c}$$

If in addition $f_o\ \varepsilon\ B_1^r$ for r>2 then we may take $\lambda\ \varepsilon\ [0,2]$ and f(t) fulfills also:

$$\int |v|^2\cdot f(v,t)\,dv = \int |v|^2\cdot f_o(v)\,dv, \quad t>0. \tag{2.12}$$

■

<u>Theorem 3</u>

If (2.9) is satisfied with $\lambda\varepsilon[0,2)$ and $f_o>0$, $f_o\varepsilon B_1^r$ for some r>4 then there exists a unique strong (in L^1) solution f(t)>0 of (2.1) for all t>0 such that (2.10), (2.11a), (2.11b) and (2.12) hold. If in addition $f_o\ln f_o\varepsilon L^1$ then

$$f(t)\ln f(t)\ \varepsilon\ L^1,\ t>0 \tag{2.13}$$

and $H(t) = \int f(v,t)\ln f(v,t)dv$ is a non-increasing function of t.

■

<u>Remark</u>

Arkeryd's theorems deliver solutions as functions $f:[0,+\infty)\to L^1$ (in Theorem 3 with derivatives in the sense of the calculus in the Banach space L^1). However any solution from Theorem 2 and Theorem 3 is a continuously differentiable function $f(v,\cdot):[0,+\infty)\to[0,+\infty)$ for a.e. $v\varepsilon R^3$ satisfying (2.1) pointwise (i.e. in the classical sense). For the discussion see C.Truesdell and R.G.Muncaster [31].

<u>Proof of Theorem 2</u>

1) k bounded.

In that case $\|J(f,g)\|_1 < c\|f\|_1\|g\|_1$. Hence there exists a unique solution f(t) of (2.1) for $t\ \varepsilon\ [0,t_1]$, where t_1 depends on k and $\|f_o\|_1$. This solution satisfies (2.11a). Suppose the solution is non-negative for every initial data $f_0>0$ then we can obtain a unique solution for $t\varepsilon[t_1,2t_1]$ (with initial data $f(t_1)$) and by induction for all $t>0$. The positivity of the solution is obtained, as usually, by appropriate successive approximation. Next if $f_0\varepsilon B_1^2$ then $f(t)\varepsilon B_1^2$ and fulfills (2.11a), (2.11b) and (2.12). If in addition $f_0\ln f_0\varepsilon L^1$ then $f(t)\ln f(t)\varepsilon L^1$ and $H(t)$ is a non-increasing function of t.

2) k unbounded.

A solution is found as the weak limit in L^1 of the sequence $\{f_n\}$ of the solutions of (2.1) with k replaced by $k_n = \min(k,n)$. For this the following lemma (see Morgenstern [25]) is applied:

<u>Lemma 3</u>

Let $\{f_n\}$ be a sequence of functions such that $f_n>0$, $f_n\varepsilon L^1$, $B_1^r\{f_n\} < c(r) <+\infty$ for some $r>0$ and such that $\int f_n(v)\ln f_n(v)dv< \text{const}<+\infty$ for all $n=1,2,...$

If $\Psi\varepsilon B_1^{r'}$ $(0<r'<r)$ then $\{f_n\}$ contains a sub-sequence $\{f_{n(j)}\}$ converging weakly to a function $f \varepsilon L^1$ and

$$\lim_j \int f_{n(j)}(v)\Psi(v)dv = \int f(v)\Psi(v)dv.$$

<u>Proof of Theorem 3</u>

<u>Existence</u>

The proof is based on a monotonicity argument. Let J_n be the collision operator with k replaced by $k_n = \min(k,n)$. J_n is not positive nor monotone hence the following initial value problems have been considered:

$$df/dt + f\cdot h(f_0) = J_n^{(i)}(f,f),\ i=1,2$$

$$f(v,0) = f_0(v), \qquad (2.14)$$

where

$$h(f)(v)=b(1+|v|^2)\iiint(1+|v_1|^2f(v_1)d\varepsilon d\theta dv_1,\ \text{b-constant}$$

and

$$J_n^{(1)}(f,f) = Q_n(f,f) + f\cdot h(f) - f\cdot P_n(f),$$

$$J_n^{(2)}(f,f) = Q_n(f,f) + f\cdot h(f) - f\cdot P(f),$$

where

$$J(f,f) = Q(f,f) - f\cdot P(f),$$

$$J_n(f,f) = Q_n(f,f) - f\cdot P_n(f).$$

For b sufficiently large $J_n^{(1)}$ and $J_n^{(2)}$ are already positive and monotone i.e. if $0<f<g$ then $0<J_n^{(i)}(f,f) < J_n^{(i)}(g,g)$ $(i=1,2)$. Moreover $J_n^{(1)}(f,f) > J_n^{(2)}(f,f)$. There exist solutions $f_n^{(i)}$ of the problem (2.14) $(i=1,2)$ and

$$f(t) = \lim_n f_n^{(2)}(t)$$

is a solution of (2.1) provided

$$\int(1+|v|^2)f(v,t)dv = \int(1+|v|^2)f_0(v)dv \qquad (2.15)$$

which can be proved by integration of (2.1) multiplied by $(1+|v|^2)$.

<u>Uniqueness</u>

Let f be an iterative solution, and g another solution of the same class then $f<g$ and

$$\int(1+|v|^2)f(v,t)dv = \int(1+|v|^2)g(v,t)dv = \int(1+|v|^2)f_0(v)dv$$

Then $f=g$ a.e.

■

Having a solution for all times we can analyze the time evolution of the solution and the relaxation of the time dependent solution to a stationary one. In the spatially uniform case there is a good candidate for the stationary solution, namely the Maxwell distribution:

$$\omega = m/(2\pi T)^{-3/2} \exp\,(-|v-u|^2/2T). \qquad (2.16)$$

It is easy to check that this is really a solution, and the only remaining question is its uniqueness. Carleman [6] has shown uniqueness in $C^{0,r}$ and Arkeryd [1] proved the uniqueness of the Maxwellian distribution as a stationary solution among all L^1 - solutions, provided the collision kernel k is a positive function almost everywhere.

Also convergence to equilibrium was investigated in the Carleman and Maslova-Tchubenko cases. Carleman [6] and then Maslova and Tchubenko [20,23] established the trend of their solutions to the equilibrium distribution function i.e. uniform convergence to the Maxwellian $\omega(v)$ with the same hydrodynamic moments as the initial data i.e.

$$m=\int f_0(v)dv,\ u=1/m\int v\cdot f_0(v)dv,\ T=1/3m\int (v-u)^2 f_0(v)dv. \quad (2.17)$$

In the L^1 case, Arkeryd [1] proved weak (in L^1) convergence of his solution to the Maxwellian (2.16) with the same hydrodynamic moments as the initial data. To have a strong convergence, as Elmroth [11] has shown, it is sufficient to prove that $\int f(v,t)\ln f(v,t)dv$ converges towards $\int\omega(v)\ln\omega(v)dv$ when t tends to infinity, where $\omega(v)$ is given by (2.16) with hydrodynamic moments (2.17). (Recently Arkeryd [4] has proved that convergence is in fact strong.)

All previously mentioned results are valid for inverse power law potentials with exponents $n>4$ and a cut-off. The problem of soft potentials i.e. with $2<n<4$, was for a long time unsolved and what is more important this was also the case for inverse power law potentials without a cut-off. This last problem is of great physical importance as most calculations made in the kinetic theory refer to intermolecular forces of infinite range. The existence problem for soft as well as hard intermolecular potentials without a cut-off was solved by Arkeryd [2] but the question of uniqueness remains open in both cases of soft potentials and potentials without a cut-off.

For forces of infinite range Arkeryd used the following weak form of the Boltzmann equation:

$$\int f(v,t)g(v,t)dv=\int f_0(v)g(v,0)dv + \int_0^t\int f(v,s)\partial g/\partial s(v,s)dvds + \int_0^t\int J(f,f)(v,s)g(v,s)dvds, \quad (2.18)$$

where test functions $g \in C^1([0,+\infty)\times R^3)$ and

$$\sup_{v,t} |g|,\ \sup_{v,t} |\partial g/\partial t|,\ \sup_{v,t} |\mathrm{grad}_v g| < +\infty.$$

To obtain (2.18) we multiply the Boltzmann equation by a test function g, integrate in t and v and carry out an integration by parts in t.

Let us assume now that k satisfies:

$$\int_0^{\pi/2} (\pi/2-\theta)k(|v-v_1|,\theta)|v-v_1|d\theta < c(1+|v|^\lambda)(1+|v_1|^\lambda) \quad (2.19)$$

for $\lambda \in [0,2]$. (This includes inverse n-th power potentials with n>2 without cut-off.) Then the following theorem holds:

<u>Theorem 4</u>

Let (2.19) be satisfied for some $\lambda\in[0,2]$ and $f_0>0$, $f_0 \in B_1^2$, $f_0 \ln f_0 \in L^1$. Then there exists a non-negative solution of (2.18) and it satisfies (2.11a), (2.11b) and (2.11c).

∎

The proof is based on a weak L^1 compactness argument (Lemma 3) and a result from the cut-off case (point 1 of the proof of Theorem 2).

Arkeryd [2] also showed for the case of soft potentials that higher moments exist for all time if they exist at t=0 and that (2.12) is satisfied. Elmroth [10] proved for hard potentials that Arkeryd's solutions have globally bounded higher moments and showed that (2.12) is satisfied provided $f_0 \in B_1^r$ for r>2. As in the cut-off case, the L^1- weak convergence towards equilibrium was established for hard potentials by Arkeryd [3](non-standard arguments) and by Elmroth [11](standard proof).

3.LOCAL EXISTENCE FOR SPATIALLY DEPENDENT EQUATION

The first existence theorem for the spatially dependent problem was due to Grad [14]. He assumed Maxwell molecules and proved the existence of solutions in a small time interval for initial data bounded by a Maxwellian. But the most fruitful local existence result was proved by Grad [16] in 1965. He considered the equation for cut-off hard potentials and initial data which are close to an equilibrium. Hence, assuming a solution in the form:

$$f = \omega_0 + \omega_0^{1/2}F,$$

where

$$\omega_0(v) = 1/(2\pi)^{3/2} \exp(-v^2/2)$$

we obtain the equation satisfied by F:

$$\partial F/\partial t + v\cdot \mathrm{grad}_x F = LF + \nu\Gamma(F,F), \tag{3.1}$$

where

$$\nu\Gamma(F,F) = \omega_0^{-1/2} J(\omega_0^{1/2} F, \omega_0^{1/2} F)$$

and L is the same as in Section 1 with ω replaced by ω_0.

Let us introduce now functional spaces in which this equation will be solved. Let $W_p^m(\Omega)$ be the usual Sobolev space. We shall consider functions which are in W_p^m with respect to the x-variable, and in $L_q(R^3)$ with some polynomial weight with respect to the v-variable. Let us denote such spaces by $B_{q,p}^{r,m}$ i.e.

$$B_{q,p}^{r,m} = \{F(x,v): x\in\Omega, v\in R^3, N_{q,p}^{r,m}\{F\} < \infty\}$$

where $N_{q,p}^{r,m}$ is the norm in $B_{q,p}^{r,m}$ given by

$$N_{q,p}^{r,m}\{F\} = \left(\int_{R^3}(1+|v|^2)^{r/2}\left(\sum_{|k|=0}^{m}\int_\Omega |D_x^k F|^p dx\right)^{q/p} dv\right)^{1/q}$$

and

$$D_x^k = \frac{\partial^{|k|}}{\partial x_1^{k_1}\,\partial x_2^{k_2}\,\partial x_3^{k_3}}, \quad |k| = k_1+k_2+k_3,$$

(extension to $p=\infty$ or $q=\infty$ is obvious).

In some cases we need more restrictions on the initial data. To formulate the restriction let us note that the operator $-v\cdot\mathrm{grad}_x F + LF$ has in $B_{q,p}^{r,m}$ a fivefold degenerated eigenvalue $\lambda=0$ and let us denote by Π the projection of $B_{q,p}^{r,m}$ on the eigenspace corresponding to this eigenvalue.

Grad considered (3.1) in a bounded rectangular domain Ω with specular boundary conditions. By reflection of the fundamental domain Ω with respect to each of three coordinate planes we obtain a domain Ω^* consisting of eight replicas of Ω. In Ω^* the function f satisfies a periodic boundary condition, hence by periodicity it can be extended to the whole

R^3 - space as a periodic function. The same is true for F and a boundary value problem for (3.1) in the case of specular reflection in a rectangular domain can be formulated as an initial value problem in a subspace of periodic functions:

$$\partial F/\partial t+v\cdot \mathrm{grad}_x\, F=LF+\nu\Gamma(F,F),\ x\in R^3, v\in R^3, t>0, \tag{3.2a}$$

$$F(x,v,0) = F_0(x,v), \tag{3.2b}$$

$$F \text{ - periodic in } x. \tag{3.2c}$$

Then the result of Grad is:

<u>Theorem 5</u>

Let $F_0 \in B_{\infty,2}^{3,3}$ and $N_{\infty,2}^{3,3}\{F_0\}$ be small enough, then the problem (3.2) has a unique solution $F(t)$ for $t<t_0$, where t_0 depends on the initial norm $N_{\infty,2}^{3,3}\{F_0\}$.

■

The proof of this theorem is based on the following lemmas:

<u>Lemma 4</u>

There is a unique solution to

$$\partial F/\partial t + v\cdot \mathrm{grad}_x\, F = LF + \nu g, \tag{3.3a}$$

$$F(0) = F_0, \tag{3.3b}$$

$$F \text{ - periodic in } x, \tag{3.3c}$$

for $N_{\infty,2}^{r,m}\{F_0\}$ and $N_{\infty,2}^{r,m}\{g\}$ finite and $r\geq 3$, which satisfies:

$$N_{\infty,2}^{r,m}\{F\}<cN_{\infty,2}^{r,m}\{F_0\}+c_1(1+t_0) \sup_{0\leq t\leq t_0} N_{\infty,2}^{r,m}\{g(t)\}. \tag{3.4}$$

<u>Lemma 5</u>

For $r\geq 0$ and $m\geq 3$ the following estimate holds:

$$N_{\infty,2}^{r,m}\{\Gamma(F,F)\} < \alpha\ (N_{\infty,2}^{r,m}\{F\})^2. \tag{3.5}$$

<u>Proof of Theorem 5</u>

We look for a solution of (3.2a) by the iteration of the nonlinear term:

$$\partial F^{(n)}/\partial t + v\cdot \mathrm{grad}_x F^{(n)} = LF^{(n)} + \nu\Gamma(F^{(n-1)},F^{(n-1)}).$$

Then, by above lemmas , we obtain:

$$N_{\infty,2}^{3,3}\{F^{(n)}\}<c_2N_{\infty,2}^{3,3}\{F_0\}+c_3 \sup_{0\leq t\leq t_0} (N_{\infty,2}^{3,3}\{F^{(n-1)}\})^2,$$

where $c_3 = c_1(1 + t_0)\alpha$.

Choosing F_0 such that

$$4c_2c_3N_{\infty,2}^{3,3}\{F_0\} < 1,$$

we obtain a bounded sequence of iterates:

$$N_{\infty,2}^{3,3}\{F^{(n)}\} < 1/2c_3.$$

Taking differences we have:

$$N_{\infty,2}^{3,3}\{F^{(n+1)}-F^{(n)}\} <$$
$$< c_3\ N_{\infty,2}^{3,3}\{F^{(n)}+F^{(n-1)}\}\cdot N_{\infty,2}^{3,3}\{F^{(n)}-F^{(n-1)}\} <$$
$$< N_{\infty,2}^{3,3}\{F^{(n)}-F^{(n-1)}\}$$

which shows that the iterations contract.

∎

As it is seen from the above proof essential difficulties are hidden in the Lemmas. We are now going to present the proof of the first of these Lemmas, which is less technical and utilizes several properties of the linearized equation.

Proof of Lemma 4

Our aim is to solve the linear nonhomogeneous problem (3.3). To this end let us multiply (3.3a) by F and integrate with respect to x and v to obtain:

$$N_{2,2}^{0,m}\{F\} < N_{2,2}^{0,m}\{F_0\} + \int_0^{t_0} N_{2,2}^{0,m}\{\nu g\}$$
$$< N_{2,2}^{0,m}\{F_0\} + ct_0 \sup_{0\le t\le t_0} N_{\infty,2}^{3,m}\{g\}, \qquad (3.6)$$

where we make use of periodicity conditions of F and nonpositivity of L.

To obtain supremum estimates we rewrite (3.3a) in the integral form:

$$F(x,v,t) = F_0(x-tv,v)\exp(-\nu t) +$$
$$\int_0^t \exp(-\nu(t-s))[KF+\nu g](x-sv,v,s)ds. \qquad (3.7)$$

From this equation we obtain:

$$N_{\infty,2}^{r,m}\{F\} < N_{\infty,2}^{r,m}\{F_0\} +$$
$$\sup_{0\le t\le t_0} (N_{\infty,2}^{r,m}\{KF\}+N_{\infty,2}^{r,m}\{g\}). \qquad (3.8)$$

Essential properties of the linearized collision operator

are:

$$N_{\infty,2}^{r,m}\{KF\} < c\, N_{\infty,2}^{r-1,m}\{F\}, \qquad (3.9)$$

$$N_{\infty,2}^{0,m}\{KF\} < c\, N_{2,2}^{0,m}\{F\}. \qquad (3.10)$$

By (3.9) and (3.10) we obtain from (3.8):

$$N_{\infty,2}^{0,m}\{F\}<N_{\infty,2}^{0,m}\{F_0\}+c \sup_{0\le t\le t_0} (N_{2,2}^{0,m}\{F\}+N_{\infty,2}^{0,m}\{g\}),$$

$$N_{\infty,2}^{r,m}\{F\}<N_{\infty,2}^{r,m}\{F_0\}+c \sup_{0\le t\le t_0} (N_{\infty,2}^{r-1,m}\{F\}+N_{\infty,2}^{r,m}\{g\}).$$

Iterating these inequalities and using (3.6) we obtain the assertion of the lemma.

∎

Another approach to the existence problem was proposed by Kaniel and Shinbrot [19]. They considered a boundary value problem in a domain $V \subset R^3$ for cut-off potentials. To formulate the results of Kaniel and Shinbrot let us rewrite (1.1) in the form:

$$\partial f/\partial t + v\cdot \mathrm{grad}_x f = Q(f,f)-fR(f), \qquad (3.11)$$

where

$$Q(f,g)=1/2\iiint k(|v-v_1|,\theta)[f(v_1')g(v')+f(v')g(v_1')]dv_1 d\varepsilon d\theta \qquad (3.12)$$

$$R(f) = \iiint k(|v-v_1|,\theta)f(v_1)dv_1 d\varepsilon d\theta. \qquad (3.13)$$

We now say a word about boundary conditions. A mapping $A{:}\,\partial V \times R^3 \to R^3$ is called a reflection law if, for every $x\in\partial V$, $n(x)\cdot v<0$ implies $n(x)A(x,v)>0$, where $n(x)$ is the inner normal to ∂V at x. Kaniel and Shinbrot introduced so called regular reflection laws for which boundary conditions have the simple form:

$$f(x,A(x,v),t+0)=f(x,v,t-0) \text{ for } x\in\partial V \text{ and } n(x)\cdot v<0. \qquad (3.14)$$

A boundary value problem for (3.11) can be written using the idea of a trajectory. A trajectory is a family of characteristics of (3.11) matched by means of reflection, i.e. it is a curve along which a gas particle which does not suffer collisions with other particles moves. Let $T_t(x_0,v_0)$ be a trajectory which pass through the point (x_0,v_0) i.e. $T_t(x_0,v_0)$ is a point (x,v) at which (x_0,v_0) has arrived by time t. Introducing the function:

$$f^{\#}(x,v,t) = f(T_t(x,v),t) \quad (3.15)$$

we can write equation (3.11) as follows:

$$\partial f^{\#}/\partial t = Q^{\#}(f,f) - f^{\#}R^{\#}(f). \quad (3.16)$$

Using (3.15) and the definition of a trajectory we observe that (3.14) has the form:

$$f^{\#}(x,v,t+0) = f^{\#}(x,v,t-0) \quad (3.17)$$

i.e. $f^{\#}$ is continuous in t on a boundary. Hence a boundary value problem for (3.11) can be reduced to the following initial value problem:

$$\partial f^{\#}/\partial t + f^{\#}R^{\#}(f) = Q^{\#}(f,f),$$
$$f^{\#}(0) = \varphi. \quad (3.18)$$

We will solve equation (3.18) by the following iterative scheme:

$$\partial l_n^{\#}/\partial t + l_n^{\#}R^{\#}(u_{n-1}) = Q^{\#}(l_{n-1},l_{n-1}), \quad (3.19a)$$
$$\partial u_n^{\#}/\partial t + u_n^{\#}R^{\#}(l_{n-1}) = Q^{\#}(u_{n-1},u_{n-1}), \quad (3.19b)$$
$$l_n(0) = u_n(0) = \varphi. \quad (3.19c)$$

To begin iteration we need a pair of functions (l_0,u_0). Following Kaniel and Shinbrot we say that such a pair satisfies the beginning condition in $[0,T]$ if:

1. $u_0 \exp(-r|v|^2) \in L^{\infty}([0,T],B_{\infty,\infty}^{0,0})$ for some $r>0$,
2. $0 \le l_0(t) \le l_1(t) \le u_1(t) \le u_0(t)$, $t \in [0,T]$.

<u>Theorem 6</u>

Let $\varphi\exp(-r|v|^2) \in B_{\infty,\infty}^{0,0}$, $\varphi \ge 0$. Then there exists T such that (3.18) possess a mild solution on $[0,T]$.

■

The proof of this theorem follows from two lemmas:

<u>Lemma 6</u>

Let $\varphi\exp(-r|v|^2) \in B_{\infty,\infty}^{0,0}$, $\varphi \ge 0$ and (l_0,u_0) satisfy the beginning condition in $[0,T]$. Then the sequence of iterative solutions of (3.19) converges to a mild solution of (3.18).

<u>Lemma 7</u>

Let $\varphi\exp(-r|v|^2) \in B_{\infty,\infty}^{0,0}$, $\varphi \ge 0$. Then there exists T such that the initial value problem:

$$\partial_t u^{\#} = Q^{\#}(u,u),$$
$$u(0) = \varphi, \tag{3.20}$$

possesses a solution u on [0,T] such that:

$$u \exp(-r|v|^2) \in L^{\infty}([0,T], B_{\infty,\infty}^{0,0}).$$

■

Proof of Theorem 6

Let u(t) be a solution from Lemma 7. Taking

$$l_0(t) = 0, \; u_0(t) = u(t)$$

and constructing $l_1(t)$, $u_1(t)$ due to (3.19) we see that

$$0 < l_1(t) < u_1(t) = u_0(t).$$

Hence (l_0,u_0) satisfy the beginning condition on [0,T] and the rest of the proof follows from Lemma 6.

■

Proof of Lemma 6

First, we prove that the sequence of recursive solutions of (3.19) satisfy:

$$0<l_0(t)<l_1(t)<\ldots<l_k(t)<\ldots<u_k(t)<\ldots<u_1(t)=u_0(t). \tag{3.21}$$

The proof of (3.21) is inductive. Suppose that $l_0, l_1, \ldots, l_{k-1}$ and $u_0, u_1, \ldots, u_{k-1}$ exist and satisfy:

$$0 < l_0 < l_1 < \ldots < l_{k-1} < \ldots < u_{k-1} < \ldots < u_1 \le u_0.$$

From (3.19) we have

$$l_k^{\#} = \varphi \exp[-\int_0^t R^{\#}(u_{k-1})] + \int_0^t Q^{\#}(l_{k-1},l_{k-1}) \exp[-\int_s^t R^{\#}(u_{k-1})]ds$$
$$\tag{3.22}$$
$$u_k^{\#} = \varphi \exp[-\int_0^t R^{\#}(l_{k-1})] + \int_0^t Q^{\#}(u_{k-1},u_{k-1}) \exp[-\int_s^t R^{\#}(l_{k-1})]ds$$

Because of the monotonicity of operators R and Q we have:

$$R^{\#}(l_{k-2}) < R^{\#}(l_{k-1}) < R^{\#}(u_{k-1}) < R^{\#}(u_{k-2}),$$

$$Q^{\#}(l_{k-2},l_{k-2})<Q^{\#}(l_{k-1},l_{k-1})<Q^{\#}(u_{k-1},u_{k-1})<Q^{\#}(u_{k-2},u_{k-2}).$$

Hence $l_{k-1} < l_k < u_k < u_{k-1}$ which proves (3.21).

Because of the Levy property the sequence $\{u_k(t)\}$ and $\{l_k(t)\}$ converge to u(t) and l(t), respectively. Subtracting l_k from u_k in (3.22) and tending with k to infinity we obtain ($\|\cdot\|$ is the L^1 norm):

$$\|u(t)-l(t)\| = \int_0^t \|Q(u(s)-l(s),u(s)+l(s))\|ds. \qquad (3.23)$$

Taking into account that $u(t)>l(t)$ and making simple estimates we obtain $u(t)=l(t)$ which proves the lemma.

■

We omit in our presentation the proof of Lemma 7 which is purely technical. Let us remark, however, that these is the lemma which makes Theorem 6 local, because Lemma 6 holds globally.

4.GLOBAL EXISTENCE FOR SPATIALLY DEPENDENT EQUATION

Grad's existence theorem from the previous section provides the existence of solutions on a time interval $[0,t_0]$ with t_0 increasing to infinity as initial data F_0 approach zero. Hence, proving that a solution of linearized problem decays to zero as t goes to infinity, we will be able to prove the global existence result for the nonlinear problem. In fact, it is the line along which first global existence theorems have been proved.

The first global result was obtained by Ukai [32] who considered the periodic problem (3.2):

<u>Theorem 7</u>

Let the initial data $F_0 \varepsilon B_{\infty,2}^{5/2+\epsilon,3/2+\epsilon} \cap \operatorname{Ker} \Pi$ and $N_{\infty,2}^{5/2+\epsilon,3/2+\epsilon}\{F_0\}$ is small enough, then the Cauchy problem (3.2a)-(3.2c) has a unique solution F(t) globally in time such that:

$$F(t) \ \varepsilon \ L^{\infty}([0,\infty),B_{\infty,2}^{5/2+\epsilon,3/2+\epsilon}) \cap C^0([0,\infty),B_{\infty,2}^{5/2,3/2}) \cap$$
$$\cap C^1([0,\infty),B_{\infty,2}^{3/2,1/2}).$$

■

A similar result was obtained by Nishida and Imai [26] for nonperiodic initial data in R^3:

Theorem 8

Let initial data F_0 be such that the sum

$\|F_0\|_{1,2} + N_{\infty,2}^{3,3}\{F_0\}$,

where $\|\cdot\|_{1,2}$ is the norm in $L^1(x,L^2(v))$, is small enough. Then there exists a unique global solution $F(t)$ of equation (3.2) in the space $B_{\infty,2}^{3,3}$.

■

We present now a simple proof of the global existence theorem near an equilibrium. To this end let us replace the initial value problem (3.2) by the following integral equation:

$$F(t) = V(t)F_0 + \int_0^t V(t-s)\nu\Gamma(F,F)(s)ds, \tag{4.1}$$

where $V(t)$ is a semigroup generated in $L^p(\Omega\times R^3)$ by the operator

$$BF = -v\cdot \text{grad}_x F + LF \tag{4.2}$$

with

$D(B) = \{F \in L^p$: 1° $F(x-tv,v)$ is absolutely continuous with respect to t for a.e. x and v;
2° $BF \in L^p$ }.

The following lemma substitutes Lemma 4:

Lemma 8

The operator B generates in every L^p, $1\leq p<\infty$, a bounded semigroup $V(t)$ for which the following estimate holds:

$$\|V(t)F\|_p < c\,\|F\|_p. \tag{4.4}$$

In addition for $F \in \ker \mathbb{P}$ we obtain:

$$\|V(t)F\|_p < c\,\exp(-\mu t)\|F\|_p,\ \mu > 0, \tag{4.5}$$

with c dependent on μ but independent of t.

($\|\cdot\|_p$ - is the L^p-norm and $\|\cdot\|$ the L^∞-norm)

■

Remark

Lemma 8 read literally is false in L^∞, for $D(B)$ is not dense in L^∞. This difficulty can be handled by the theory of dual operators of Hille and Phillips (cf. Palczewski [27]). Then

V(t) exists as a bounded operator which satisfies (4.4) and (4.5) on a closed subspace of L^∞ in which D(B) is dense.

We will look for a global solution of equation (4.1) in the following space:

$$BP=\mathrm{cl}\{F\in L^\infty(\Omega\times R^3):(1+|v|)F\in L^\infty,(1+|v|)\mathrm{grad}_x F\in L^\infty\}, \quad (4.6)$$

where cl means closure in L^∞.

Lemma 9

The semigroup V(t) and the nonlinear operator Γ map BP into BP. ∎

This lemma assures that the integral equation (4.1) is well posed in BP. Let us consider now the following linear equation:

$$F(t) = V(t)F_0 + \int_0^t V(t-s)\nu h(s)ds. \quad (4.7)$$

Lemma 10

Let $F_0,h \in \ker \Pi\cap BP$. The linear equation (4.7) possesses in BP a unique solution for which the following estimate holds:

$$|F|_\beta \le a\|F_0\| + b|h|_\beta, \quad (4.8)$$

where

$$|F|_\beta = \sup_{t\ge 0} \exp(\beta t)\|(1+|v|)F(t)\|. \quad (4.9)$$

Proof

By (1.14) and properties of V we obtain:

$$\|(1+|v|)V(t)\nu h\| \le c\|(1+|v|)h\|. \quad (4.10)$$

Hence the right hand side of (4.7) certainly belongs to BP and it is enough to prove (4.8).

First, let us observe that for $F \in \ker \Pi$ the following estimate can be proved:

$$\|(1+|v|)V(t)F\| \le c \exp(-\beta_1 t)\|F\|, \quad (4.11)$$

where $0<\beta_1<\min(c_1,\mu)$ (cf. (1.14) and (4.5)).

Multiply (4.8) by $\exp(\beta t)(1+|v|)$ and use (4.11) to obtain:

$$|F| \le c \exp((\beta-\beta_1)t)\|F_0\| + c \exp(\beta t)\int_0^t \exp(-\beta_1(t-s))\exp(-\beta s)|h|_\beta ds. \quad (4.12)$$

Choosing $\beta<\beta_1$, we obtain the estimate (4.8).

■

Now we are able to prove our global existence theorem:

<u>Theorem 9</u>

Let $F_0 \in \ker \Pi \cap BP$ and $\|F_0\|$ be small enough. Then the initial value problem (3.2) possesses a unique mild solution $F \in BP$ globally in time.

We call F a mild solution of (3.2), for it solves in fact the integral equation (4.1).

<u>Proof</u>

We will solve the integral equation (4.1) by the method of successive approximations:

$$F^{(n)}(t) = V(t)F_0 + \int_0^t V(t-s)\nu\Gamma(F^{(n-1)},F^{(n-1)})ds. \quad (4.13)$$

By (3.5) and (4.8) the following estimate can be obtained:

$$|F^{(n)}|_\beta \le a\|F_0\| + b(|F^{(n-1)}|_\beta)^2. \quad (4.14)$$

With (4.14) the rest of the proof is analogous to the proof given by Grad (see the previous section). Hence choosing $\|F_0\|$ small enough we can made the sequence of successive approximations convergent.

The global existence results described above are obtained for initial data which are small perturbations of the Maxwell distribution that is an equilibrium solution. The question arise whether similar result can be obtained for other equilibrium solutions. A positive answer was obtained by Illner and Shinbrot [18] for a perturbation of vaccum, which is also an equilibrium solution for the Boltzmann equation.

The proof of Illner and Shinbrot is essentially based on results of Kaniel and Shinbrot described in the previous section. As we have remarked earlier what makes the result of Kaniel and Shinbrot local in time is the beginning condition. In the case of rare gas expanding to vaccum, Illner and Shinbrot found the beginning condition which is satisfied globally.

Following these authors we shall show that the inequality:

$$u^{\bullet}(t) > \varphi + \int_0^t Q^{\bullet}(u,u)ds \qquad (4.15)$$

possesses a global solution (cf. (3.20)) in a space $S_\alpha(R^3 x R^3)$, $\alpha>0$, consisting of the completion of continuous functions of compact support with respect to the norm:

$$\|f\|_\alpha = \int_{R^3} \sup_{x \varepsilon R^3} |\exp(\alpha|x|^2) f(x,v)| dv. \qquad (4.16)$$

Let

$$\chi(v) = \sup_{x \varepsilon R^3} \exp(\alpha|x|^2) f_0(x,v) \qquad (4.17)$$

and

$$\int (1+|v|)|\chi(v)| dv < +\infty. \qquad (4.18)$$

The following lemma holds:

<u>Lemma 11</u>

Let $\varphi \varepsilon S_\alpha$, $\varphi \geq 0$. If χ defined by (4.17) satisfies (4.18) and $\int|\chi(v)|dv$ is small enough then (4.15) possesses a global solution in S_α.

<u>Proof</u>

Let $u(x,v,t) = \exp(-\alpha|x-vt|^2)w(v)$. Then (4.15) can be reduced to:

$$w(v) > \chi(v) + \int_0^t \iiint k(|v-v_1|,\theta)\exp[-\alpha|x+s(v-v_1)|^2] x$$

$$x w(v')w(v_1')dv_1 d\varepsilon d\theta ds. \qquad (4.19)$$

We are looking for nonnegative solutions of (4.19). Performing integration with respect to s and using the estimate:

$$0 < k(|v-v_1|,\theta) < |v-v_1|$$

we can show that (4.19) holds if w(v) satisfies:

$$w(v) = \chi(v) + \sqrt{(\pi/\alpha)} \iiint w(v')w(v_1')dv_1 d\varepsilon d\theta. \qquad (4.20)$$

Let

$$G(w)(v) = \chi(v) + \sqrt{(\pi/\alpha)} \iiint w(v')w(v_1')dv_1 d\varepsilon d\theta. \qquad (4.21)$$

Integrating (4.21) and using nonnegativeness of w we obtain:

$$\|G(w)\| = \|\chi\| + 2\pi\sqrt{(\pi/\alpha)}\ \|w\|^2, \tag{4.22}$$

where $\|\cdot\|$ is the L^1 - norm.

Similarly we obtain:

$$\|G(w_1) - G(w_2)\| < 2\pi\sqrt{(\pi/\alpha)}(\|w_1\| + \|w_2\|)\|w_1 - w_2\|. \tag{4.23}$$

Hence, G maps the set of nonnegative functions with $\|w\| < R_0$ into itself provided:

$$\|\chi\| + 2\pi\sqrt{(\pi/\alpha)}\ R_0^2 < R_0 \tag{4.24}$$

and G is a contraction if:

$$4\pi\sqrt{(\pi/\alpha)}\ R_0 < 1. \tag{4.25}$$

Choosing χ such that:

$$8\pi\sqrt{(\pi/\alpha)}\ \|\chi\| < 1$$

we can satisfy both (4.24) and (4.25).

■

With Lemma 11 the beginning condition can be satisfied globally in time and we are able to prove the following theorem:

<u>Theorem 10</u>

Let $\varphi \in S_\alpha$, $\varphi \geq 0$. If χ defined by (4.17) satisfies (4.18) and the smallness condition $16\pi\sqrt{(\pi/\alpha)}\ \|\chi\| < 1$ then the initial value problem:

$$\partial f/\partial t + v\cdot \text{grad}_x f = J(f,f),$$

$$f(0) = \varphi \tag{4.26}$$

possesses a unique global mild solution $f(t) \in S_\alpha$.

■

To prove this theorem we need an analog of Lemma 6 in spaces S_α. This can be proved similarly as in the local case and we omit details. Let us remark, however, that condition (4.18) is essentially needed in this proof.

REFERENCES

1. Arkeryd,L. - On the Boltzmann equation, *Arch. Rat. Mech. Anal.*, <u>45</u>, 1, 1972.

2. Arkeryd,L. - Intermolecular forces of infinite range and the Boltzmann equation, *Arch. Rat. Mech. Anal.*, 77,11, 1981.
3. Arkeryd,L. - Asymptotic behaviour of the Boltzmann equation with infinite range forces, *Commun. Math. Phys.*, 86, 475, 1982.
4. Arkeryd,L. - On the Boltzmann equation in unbounded space far from equilibrium and the limit of zero mean free path, *Commun. Math. Phys.*, 105, 205, 1986.
5. Caflisch,R.E. - The Boltzmann equation with soft potential, *Commun. Math. Phys.*, 74, 71, 1980.
6. Carleman,T. - Sur la theorie de l'equation integro-differentielle de Boltzmann, *Acta Math.*, 60, 91, 1933.
7. Carleman,T. - *Probleme Mathematiques dans la Theorie Cinetique des Gas*, Upsala 1957.
8. Cercignani,C. - On Boltzmann equation with cut-off potentials, *Phys. Fluids*, 10, 2097, 1967.
9. Drange,H.B. - The linearized Boltzmann collision operator for cut-off potentials, *SIAM J. Appl.Math.*, 29, 665,1975.
10. Elmroth,T. - Global boundedness of moments of solution of the Boltzmann equation for infinite range forces, *Arch. Rat. Mech. Anal.*, 82, 1, 1983.
11. Elmroth,T. - On the H-function and convergence towards equilibrium for a space-homogeneous molecular density, *SIAM J. Appl. Math.*, 44, 150, 1984.
12. Fiszdon,W.,Lachowicz,M.,Palczewski,A.- Existence problems of the non-linear Boltzmann equation,in *Trends and Applications of Pure Mathematics to Mechanics*, Ciarlet,P.G. and Roseau,M., Eds.,Lect.Notes in Phys. No 195 Springer, Berlin 1984, p.63.
13. Gluck,P. - Solution of the Boltzmann equation, *Trans. Th. Stat. Phys.*, 9, 43, 1980.
14. Grad,H. - Principles of the kinetic theory of gases, in *Handbuch der Physik*, Fluegge,S., Ed., Springer, Berlin

1958, vol 12, 205.

15.Grad,H. - Asymptotic theory of the Boltzmann equation II, in *Proc. III Symp. RGD*, Laurmann,J.A.,Ed.,Academic Press, New York 1963, vol 1, 26.

16.Grad,H. - Asymptotic equivalence of the Navier-Stokes and nonlinear Boltzmann equation, *Proc.Symp.Appl.Math.*, 17, Amer.Math.Soc.,Providence, R.I. 1965, 154.

17.Greenberg,W.,Polewczak,J.,Zweifel,P.F. - Global existence proofs for the Boltzmann equation, in *Nonequilibrium Phenomena I: The Boltzmann Equation*, Lebowitz,J.L. and Montroll,E.W.,Eds.,North-Holland, Amsterdam 1983, p.19.

18.Illner,R.,Shinbrot,M. - The Boltzmann equation: global existence for a rare gas in an infinite vacuum, *Commun. Math. Phys.*, 95, 217, 1984.

19.Kaniel,S.,Shinbrot,M. - The Boltzmann equation: uniqueness and local existence, *Commun. Math. Phys.*, 58, 65, 1978.

20.Maslova,N.B., Tchubenko,R.P. - Asymptotic properties of solutions of the Boltzmann equation (in Russian), *Dokl. Acad. Nauk SSSR*, 202, 800, 1972.

21.Maslova,N.B., Tchubenko,R.P. - On solutions of the non-stationary Boltzmann equation (in Russian), *Vestnik Leningrad Univ.*, 1, 100, 1973.

22.Maslova,N.B.,Tchubenko,R.P. - Lower bounds of solutions of the Boltzmann equation (in Russian), *Vestnik Leningrad Univ.*, 7, 109, 1976.

23.Maslova,N.B.,Tchubenko,R.P. - Relaxation in a monoatomic space-homogeneous gas (in Russian), *Vestnik Leningrad Univ.*, 13, 90, 1976.

24.Morgenstern,D. - General existence and uniqueness proof for spatially homogeneous solutions of the Maxwell-Boltzmann equation in the case of Maxwellian molecules, *Proc. Nat. Acad. Sci. U.S.A.*, 40, 719, 1954.

25.Morgenstern,D. - Analytical studies related to the Maxwell-Boltzmann equation, *J. Rat. Mech. Anal.*, 4, 533, 1955.

26.Nishida,T.,Imai,K.- Global solutions to the initial value problem for the nonlinear Boltzmann equation, *Publ. RIMS*, 12, 229,1976.

27.Palczewski,A.- Evolution operators generated by the space and time nonhomogeneous linearized Boltzmann operator, *Trans. Th. Stat. Phys.*, 14, 1, 1985.

28.Palczewski,A. - Existence of global solutions to the Boltzmann equation in L^{∞}, in *Proc. XV Symp.RGD*, Boffi,V. and Cercignani,C.,Eds.,B.G.Teubner,Stuttgart 1986, vol.1, p.144.

29.Povzner,A.Ya. - Boltzmann equation in the kinetic theory (in Russian), *Mat. Sbornik*, 58, 65, 1962.

30.Shizuta,Y. - On the classical solutions of the Boltzmann equation, *Comm.Pure Appl.Math.*, 36, 705, 1983.

31.Truesdell,C.,Muncaster,R.G. - *Fundamentals of Maxwell's Kinetic Theory of a Simple Monoatomic Gas*, Academic Press, 1980.

32.Ukai,S. - On the existence of global solutions of mixed problem for non-linear Boltzmann equation, *Proc. Japan Acad.*, 50, 179, 1974.

ASYMPTOTICS OF THE BOLTZMANN EQUATION AND FLUID DYNAMICS*

R.E. Caflisch
New York University, New York, U.S.A.

ABSTRACT

The relations and differences between the Boltzmann equation and the fluid dynamic equations is one of the most interesting features of kinetic theory. In these four lectures at CISM in Udine I present an outline of this theory and some of its main results: derivation and validity of the fluid equations using the Hilbert and Chapman-Enskog expansions, description of shock waves, boundary layer solutions, and fluid dynamics of discrete velocity models. The objective of these lectures is to present rigorous results with solid physical meaning and to suggest further lines of research.

* The work described here was supported in part by the Air Force Office of Scientific Research through grants AFOSR-85-0017, and AFOSR-86-0352.

0. Introduction

The relations and differences between the Boltzmann equation and the fluid dynamic equations is one of the most interesting features of kinetic theory. In these four lectures at CISM in Udine I present an outline of this theory and some of its main results: derivation and validity of the fluid equations using the Hilbert and Chapman-Enskog expansions, description of shock waves, boundary layer solutions, and fluid dynamics of discrete velocity models. The objective of these lectures is to present rigorous results with solid physical meaning and to suggest further lines of research.

Two parts of these lectures are at least partly new. The Hilbert and Chapman-Enskog expansion are presented in a somewhat new and more unified way. The discussion of discrete velocity models and the possible fluid dynamic equations for such models differs from previous work, as described in that lecture, mainly through its viewpoint.

Lecture 1. The Hilbert and Chapman-Enskog Expansions

1.1 The Boltzmann Equation

The Boltzmann equation is

$$DF = \frac{1}{\varepsilon} Q(F,F) \tag{1.1}$$

in which $F = F(\underset{\sim}{x},\underset{\sim}{\xi},t)$ is the molecular distribution function, $D = \frac{\partial}{\partial t} + \underset{\sim}{\xi} \cdot \frac{\partial}{\partial \underset{\sim}{x}}$ is the convective differential operator, Q is the non-linear collision integral and ε is the (non-dimensionalized) mean free path. We shall assume that ε is very small, in which case we expect F to have the form

$$F = M + \varepsilon\tilde{F} \tag{1.2}$$

with $Q(M,M) = 0$. The only solutions of the latter equation are the local Maxwellian distributions given by

$$\begin{aligned} M &= M(\underset{\sim}{\xi};\rho_0,\underset{\sim}{u},T_0) \\ &= \rho_0(2\pi T)^{-3/2}\exp\{-|\underset{\sim}{\xi} - \underset{\sim}{u}_0|^2/2T_0\} \end{aligned} \tag{1.3}$$

in which ρ_0, $\underset{\sim}{u}_0$, T_0 may depend on $\underset{\sim}{x}$ and t. Later on ρ_0, $\underset{\sim}{u}_0$ and T_0 (and thus M) will also be allowed to depend on ε.

Under the ansatz (1.2) the Boltzmann equation (1.1) becomes

$$DM + \varepsilon D\tilde{F} = L\tilde{F} + \varepsilon Q(\tilde{F},\tilde{F}) \tag{1.4}$$

in which L is the linearized collision operator given by

$$L\tilde{F} = 2Q(M,\tilde{F}) \ . \tag{1.5}$$

The operator L is self-adjoint with respect to the inner product

$$\langle F,G\rangle = \int M^{-1} F G d\xi \qquad (1.6)$$

and it satisfies the Fredholm property

$$Rg(L) = Nu\ell(L)^{\perp} \qquad (1.7)$$

in which Rg and Nuℓ stand for range and null space and "$\perp$" means orthogonal with respect to the inner product $\langle\cdot,\cdot\rangle$.

Define projection operators P and $P^{\perp} = (1 - P)$ onto Nuℓ(L) and Rg(L) respectively, with

$$Nu\ell(L) = \mathrm{span}\{M, \xi_i M, |\xi|^2 M\}$$

$$P\tilde{F} = \{(\tilde{\rho}/\rho_0 - 3\tilde{T}/2T_0) + T_0^{-1}\, \underset{\sim}{\alpha} \cdot (\xi - \underset{\sim}{u}_0) + (\tilde{T}/2T_0^2)|\xi - \underset{\sim}{u}_0|^2\}M \qquad (1.8)$$

in which the fluid dynamic variables corresponding to $\tilde{F}$ (and linearized around M) are

$$\tilde{\rho} = \int \tilde{F} d\xi$$

$$\tilde{\underset{\sim}{u}} = \rho_0^{-1} \int (\underset{\sim}{\xi} - \underset{\sim}{u}_0) \tilde{F} d\underset{\sim}{\xi} \qquad (1.9)$$

$$\tilde{T} = \rho_0^{-1} \frac{1}{3} \int (|\underset{\sim}{\xi} - \underset{\sim}{u}_0|^2 - 3T_0) \tilde{F} d\underset{\sim}{\xi} \ .$$

In addition let Ψ and Φ denote the fluid-dynamic and non-fluid-dynamic components of $\tilde{F}$, i.e.

$$\tilde{\Phi} = P^{\perp}\tilde{F} \qquad \tilde{\Psi} = P\tilde{F} \ . \qquad (1.10)$$

Observe that the collision operator is purely non-fluid-dynamic: for any F,G

$$PQ(F,G) = 0 \ . \qquad (1.11)$$

Also if $\phi \in Rg(L)$ then

$$\langle \phi, L\phi \rangle \le -c\langle (1 + |\xi|)^{\gamma}\phi, \phi \rangle \qquad (1.12)$$

in which c and γ are constants depending on the intermolecular force law. For hard spheres $\gamma = 1$, otherwise $0 < \gamma < 1$.

Using the notation (1.10) and property (1.11), equation (1.4) (or (1.1)) is projected onto $Nu\ell(L)$ and $Rg(L)$ as

$$P^{\perp}DM + \varepsilon P^{\perp}D(\tilde{\Phi} + \tilde{\Psi}) = L\tilde{\Phi} + \varepsilon Q(\tilde{\Phi} + \tilde{\Psi}, \tilde{\Phi} + \tilde{\Psi}) \tag{1.13}$$

$$PDM + \varepsilon PD(\tilde{\Phi} + \tilde{\Psi}) = 0\ . \tag{1.14}$$

Solution of these equations will proceed by inverting L in (1.13) (which is analogous to the Lyapunov-Schmidt equation of bifurcation theory). Equation (1.14) (analogous to the bifurcation equation), which arises as a solveability condition for the inversion of L in (1.4), is a differential equation in $\underset{\sim}{x}$,t for which initial and boundary conditions can be imposed. This solution, using the smallness of ε is carried out in several different ways in the following two sections.

1.2 The Hilbert Expansion

The most straightforward expansion is that of Hilbert. Assume that M is independent of ε and that $\tilde{\Phi}$, $\tilde{\Psi}$ have expansions of the form

$$\tilde{\Phi} = \Phi_1 + \varepsilon\Phi_2 + \cdots \tag{1.15}$$

$$\tilde{\Psi} = \Psi_1 + \varepsilon\Psi_2 + \cdots\ . \tag{1.16}$$

Expand equation (1.13) as

$$L\Phi_1 = P^{|}DM \tag{1.17^1}$$

$$L\Phi_2 = P^{|}D(\Psi_1 + \Phi_1) - Q(\Phi_1 + \Psi_1, \Phi_1 + \Psi_1) \tag{1.17^2}$$

$$L\Phi_3 = P^{|}D(\Psi_2 + \Phi_2) - 2Q(\Phi_1 + \Psi_1, \Phi_2 + \Psi_2) \tag{1.17^3}$$

$$\cdots$$

In each of these equations the right hand side is in Rg(L), so that once the right hand side is known there is a unique solution $\Phi_1 \in \text{Nu}\ell(L)^{|}$. Next equation (1.14) is expanded as

$$PDM = 0 \tag{1.18^0}$$

$$PD\Psi_1 = - PD\Phi_1 \tag{1.18^1}$$

$$PD\Psi_2 = - PD\Phi_2 \tag{1.18^2}$$

$$\cdots$$

The procedure for solving (1.17^i) and (1.18^i) simultaneously is to solve first (1.18^0) for M, then (1.17^1) for Φ_1, then (1.18^1) for Ψ_1, etc.

Equation (1.18^0) is exactly the compressible Euler equations [1,3,4] for ρ_0, $\underset{\sim}{u}_0$, T_0, i.e.

$$\frac{\partial}{\partial t}\rho_0 + \frac{\partial}{\partial \underset{\sim}{x}} \cdot \rho_0 \underset{\sim}{u}_0 = 0 \tag{1.19}$$

$$\frac{\partial}{\partial t}\rho_0 \underset{\sim}{u}_0 + \frac{\partial}{\partial \underset{\sim}{x}} \cdot \rho_0 \underset{\sim}{u}_0 \underset{\sim}{u}_0 + \frac{\partial}{\partial \underset{\sim}{x}} \rho_0 T_0 = 0 \tag{1.20}$$

$$\frac{\partial}{\partial t}\rho_0(\tfrac{1}{2}|\underset{\sim}{u}_0|^2 + \tfrac{3}{2}T_0) + \frac{\partial}{\partial \underset{\sim}{x}} \cdot \rho_0 \underset{\sim}{u}_0(\tfrac{1}{2}|\underset{\sim}{u}_0|^2 + \tfrac{5}{2}T_0) = 0 \ . \tag{1.21}$$

Similarly (1.18^1) is exactly the linearized Euler equations for ρ_1, $\underset{\sim}{u}_1$, T_1 linearized around ρ_0, $\underset{\sim}{u}_0$, T_0 and with forcing terms (which depend only on ρ_0, $\underset{\sim}{u}_0$, T_0) given by

$$-PD\Phi_1 = -(PDL^{-1}P^{|}D)M \ . \tag{1.22}$$

This forcing term clearly involves second derivatives of ρ_0, $\underset{\sim}{u}_0$, T_0 and will be seen in the next section to be precisely the viscous and heat-conductive terms from the Navier-Stokes equations.

Again (1.18^2) is the linearized Euler equations for ρ_2, $\underset{\sim}{u}_2$, T_2 with forcing terms

$$\begin{aligned} -PD\Phi_2 &= -PDL^{-1}\{P^{|}D(\Psi_1 + \Phi_1) - Q(\Phi_1 + \Psi_1, \Phi_1 + \Psi_1)\} \\ &= -PD(L^{1-}P^{|}D)^2M - (PDL^{-1}P^{|}D)\Psi_1 \\ &\quad + (PDL^{-1})Q(\Phi_1 + \Psi_1, \Phi_1 + \Psi_1) \ . \end{aligned} \tag{1.23}$$

The first term in (1.23) is precisely the third order terms in ρ_0, $\underset{\sim}{u}_0$, T_0 of the Burnett equations. The second term is the linearized Navier-Stokes terms in ρ_1, $\underset{\sim}{u}_1$, T_1 and the third term is nonlinearities which were left out of the linearized Euler equations for ρ_1, $\underset{\sim}{u}_1$, T_1.

1.3 The Chapman-Enskog Expansion

The Chapman-Enskog expansion is motivated by the observation that for $M = M(\underset{\sim}{\xi};\rho_0,\underset{\sim}{u}_0,T_0)$

$$\frac{\partial}{\partial\rho} M = \rho_0^{-1} M \tag{1.24}$$

$$\frac{\partial}{\partial \underset{\sim}{u}} M = T_0^{-1}(\underset{\sim}{\xi} - \underset{\sim}{u}_0) M \tag{1.25}$$

$$\frac{\partial}{\partial T} M = \tfrac{1}{2}[|\underset{\sim}{\xi} - \underset{\sim}{u}_0|^2 - 3T_0\}T_0^{-2} M . \tag{1.26}$$

Thus if Ψ is given by (1.10) then

$$M(\rho_0,\ u_0,\ T_0) + \varepsilon\Psi = M(\rho_0 + \varepsilon\tilde{\rho},\ \underset{\sim}{u}_0 + \varepsilon\underset{\sim}{\tilde{u}},\ T_0 + \varepsilon\tilde{T}) + O(\varepsilon^2) .$$

This suggests that Ψ be eliminated by incorporating it into M; the result is the Chapman-Enskog ansatz that M depends on ε, $\Psi = 0$ and Φ is expanded as in (1.15). Then equation (1.13) becomes

$$L\Phi_1 = P^{\perp}DM \tag{1.27^1}$$

$$L\Phi_2 = P^{\perp}D\Phi_1 - Q(\Phi_1,\Phi_1) \tag{1.27^2}$$

$$L\Phi_3 = P^{\perp}D\Phi_2 - 2Q(\Phi_1,\Phi_2) \tag{1.27^3}$$

$$\cdots$$

and equation (1.14) becomes

$$PDM = -PD(\varepsilon\Phi_1 + \varepsilon^2\Phi_2 + \varepsilon^3\Phi_3 + \cdots) \ . \tag{1.28}$$

Since M (which depends on ε) is not written as an expansion in powers of ε (as it was in the Hilbert expansion), equation (1.28) cannot be expanded. However this equation can be approximated at successively higher order as

$$PDM = 0 \tag{1.29^0}$$

$$PDM = -\varepsilon PD\Phi_1 \tag{1.29^1}$$

$$PDM = -\varepsilon PD\Phi_1 - \varepsilon^2 PD\Phi_2 \tag{1.29^2}$$

The error in (1.29^i) as an approximation to (1.28) is $O(\varepsilon^{i+1})$. Equation (1.29^0) is precisely the non-linear Euler equations (1.19) - (1.21); equations (1.29^1) is the non-linear Navier-Stokes equations [1,3,4] for ρ_0, u_0, T_0 given by (in one space dimension)

$$\frac{\partial}{\partial t} \rho_0 + \frac{\partial}{\partial x} \rho_0 u_0 = 0 \tag{1.30}$$

$$\frac{\partial}{\partial t} \rho_0 u_0 + \frac{\partial}{\partial x} (\rho_0 u_0 u_0) + \frac{\partial}{\partial x} \rho_0 T_0 = \varepsilon \frac{\partial}{\partial x} \{\mu(T_0) \frac{\partial}{\partial x} u_0\} \tag{1.31}$$

$$\frac{\partial}{\partial t} \rho_0(\frac{1}{2}u_0^2 + \frac{3}{2}T_0) + \frac{\partial}{\partial x} \rho_0 u_0(\frac{1}{2}u_0^2 + \frac{5}{2}T_0) \tag{1.32}$$

$$= \varepsilon \frac{\partial}{\partial x} \{\eta(T_0) \frac{\partial}{\partial x} T_0\} + \varepsilon \frac{\partial}{\partial x} \{\mu(T_0) u_0 \frac{\partial}{\partial x} u_0\} .$$

The terms $\varepsilon PD\Phi_1$ is the second order viscosity and heat-conductivity terms as stated in the previous section. Similarly (1.29^2) is the nonlinear Burnett system with third order terms given by $-\varepsilon^2 PD\Phi_2$.

As Grad pointed out [5], the role of the approximate equations (1.29^i) may be summarized by saying that the Chapman-Enskog expansion is an expansion of the Boltzmann equation rather than an expansion of its solution.

1.4 Validity of the Expansions

These expansions are expected to be asymptotically valid for small ε in the regular regions of the solutions; i.e. away from initial layers, boundary layers and shocks. This result is stated in the following theorem [7,8,2].

Theorem 1.1. Let $(\rho_0(\underset{\sim}{x},t), \underset{\sim}{u}_0(x,t), T_0(\underset{\sim}{x},t))$ be a smooth spatially periodic solution of the nonlinear Euler equations (1.19) - (1.21), (with $\rho_0 > 0$, $T_0 > 0$) for $t \in [0,\tau]$, $\underset{\sim}{x} \in [0,1]^3$. Let $F_0(\underset{\sim}{x},\xi)$ be a distribution function with $||M(\rho_0,\underset{\sim}{u}_0,T_0) - F_0||$ sufficiently small. Then for $0 \leq t \leq \tau$ the Boltzmann equation (1.1) has a smooth solution F^ε with initial data F_0. Moreover there is a corresponding truncated Hilbert expansion $F_n^\varepsilon = M + (\varepsilon\Phi_1 + \cdots + \varepsilon^n\Phi_n) + (\varepsilon\Psi_1 + \cdots + \varepsilon^n\Psi_n)$ solving equations (1.17^i) and (1.18^i). For any $0 < \tau_0 < \tau$ it satisfies

$$\sup_{\tau_0 < t \leq \tau} ||F^\varepsilon - F_n^\varepsilon|| \leq c\varepsilon^{n+1} \tag{1.33}$$

in which the constant c depends only on τ_0.

Note: 1) The assumption of periodicity is just a technical simplification. derivatives.

2) The norm in (1.33) is a weighted sup norm in velocity and a square-integral Sobolev norm in space.

There are corresponding results for the Chapman-Enskog expansion. There are also results which show that the Navier-Stokes equations give the correct long-time asymptotics of the Boltzmann solution [6].

1.5 References

[1] R. Caflisch, Fluid dynamics and the Boltzmann equation, in Nonequilibrium Phenomenon I: The Boltzmann Equation, eds. Montroll and Lebowitz, North-Holland (1983) 193-223.

[2] R. Caflisch, The fluid dynamic limit of the nonlinear Boltzmann equation. Comm. Pure Appl. Math. 33 (1980) 651-666.

[3] C. Cercignani, Theory and Application of the Boltzmann equation (Elsevier, Amsterdam, 1975).

[4] S. Chapman and T.G. Cowling. The Mathematical Theory of Nonuniform Gases, 3d Ed. (Cambridge Univ. Press, Cambridge, 1970).

[5] H. Grad, Principles of the kinetic theory of gases. Handbuch der Physik, Vol. 12 (Springer, Berlin, 1958) pp. 205-294.

[6] S. Kawashima, A. Matsumara and T. Nishida. On the fluid-dynamical approximation to the Boltzmann equation at the level of the Navier-Stokes equation. Commun. Math. Phys. 70 (1979) 97-124.

[7] M. Lachowicz, On the initial layer and the existence theorm for the nonlinear Boltzmann equation. Math. Meth. in Appl. Sci. to appear.

[8] M. Lachowicz, On the initial layer and the existence theorem for the nonlinear Boltzmann equation, Part II: Differentiality of

the solution of some linear system of equations. Archives of Mechanics. to appear.

Lecture 2. Shock Wave Solutions

2.1 Shock Equations and Jump Conditions

A plane steady shock profile is a continuous solution $F(\xi,x,t) = F(\xi,x-st)$ which depends on only one space variable $x = x_1$ and translates at uniform speed s. Its values at $x = \pm\infty$ are Maxwellians $M_\pm$ given by (1.3) with $\rho_\pm, u_\pm, T_\pm$ each constant and $\underset{\sim}{u}_\pm = (u_\pm,0,0)$. Therefore the Boltzmann equation (1.1) becomes (with $\varepsilon = 1$)

$$(\xi_1 - s)\frac{\partial}{\partial x} F = Q(F,F) \qquad (2.1)$$

in which ξ_1 is the first component ξ, and there are asymptotic conditions

$$F(\xi,x = \pm\infty) = M_\pm(\xi)$$
$$= \rho_\pm(2\pi T_\pm)^{-3/2}\exp\{-((\xi_1-u_\pm)^2 + \xi_2^2 + \xi_3^2)/2T_\pm\} \qquad (2.2)$$

The conservation properties (1.11) imply that

$$\frac{\partial}{\partial x}\int(\xi_1-s)Fd\underset{\sim}{\xi} = 0$$

$$\frac{\partial}{\partial x}\int\xi_1(\xi_1-s)Fd\underset{\sim}{\xi} = 0 \qquad (2.3)$$

$$\frac{\partial}{\partial x}\int\xi^2(\xi_1-s)Fd\underset{\sim}{\xi} = 0$$

so that the quantities in brackets are constant in x. Equating their values at $x = \pm\infty$ and using the form of $F(x = \pm\infty)$ results in the Rankine-Hugoniot conditions:

$$(u_- - s)\rho_- = (u_+ - s)\rho_+$$

$$(u_- - s)^2\rho_- + \rho_-T_- = (u_+ - s)^2\rho + \rho_+T_+$$

$$(u_- - s)\rho_-(\frac{5}{2}T_- + \frac{1}{2}(u_--s)^2)$$

$$= (u_+ - s)\rho_+(\frac{5}{2}T_+ + \frac{1}{2}(u_+-s)^2) \ . \qquad (2.4)$$

The Boltzmann collison operator Q satisfies $\int \log F\, Q(F,F)d\underset{\sim}{\xi} \le 0$. Use of this inequality with (2.1) and (2.3) results in

$$\frac{\partial}{\partial x}\int (\xi_1 - s)F \log F \, d\underset{\sim}{\xi} \le 0 \ . \qquad (2.5)$$

This is the analogue of the Boltzmann H-theorem for the shock problem.

If we compare the values of this integral at $x = \pm\infty$ as before, we obtain the entropy inequality:

$$\mathrm{sgn}(u_- - s)S_- < \mathrm{sgn}(u_+ - s)S_+ \qquad (2.6)$$

in which $S = \log(\rho^{-2/3}T)$ is the fluid dynamic entropy. Note that $\mathrm{sgn}(u_- - s) = \mathrm{sgn}(u_+ - s)$.

The Rankine-Hugoniot conditions (2.4) and the entropy inequality (2.6) are exactly the description from the Euler equations of a steady planar shock. This shows the agreement between the Boltzmann theory and the Euler theory. Note that although (2.4) and (2.6) are relations only for the asymptotic values of F, the corresponding relations (2.3) and (2.5) apply for all x.

General solutions for the shock profile problem are quite difficult. The approximate method of Mott-Smith [4] and Tamm [5] has proved to be very successful in practice, but has not received a mathematical basis. It has been modified for infinitely strong shocks in [3]. Some preliminary mathematical results on infinitely strong shocks are presented in [2].

2.2 Weak Shocks

Suppose further that $M_+ = M_- + O(\varepsilon)$, i.e. the shock is weak. Write

$$F = M_-(\xi) + \varepsilon f(y = \varepsilon x, \xi) \qquad (2.7)$$

satisfying

$$(\xi_1 - s)\frac{\partial}{\partial y} f = \frac{1}{\varepsilon} Lf + Q(f,f) \tag{2.8}$$

$$f(-\infty) = 0 \qquad f(+\infty) = M_+ - M_-$$

in which $Lf = 2Q(M_-,f)$. We shall find the dominant terms in f. Note that now ε is the shock strength; the mean free path is set to 1.

For fixed M_-, let the possible equilibria states be parameterized by ε, i.e. $M_+ = M_+(\varepsilon)$, with $M_+(\varepsilon = 0) = M$ or

$$M_+(\varepsilon) = M_- + \varepsilon\Psi_- + O(\varepsilon^2) \ . \tag{2.9}$$

Also the shock speed s depends on ε. We first find $s_0 = s(\varepsilon = 0)$ and then find Ψ_0. The R-H conditions say that

$$<(\xi_1 - s(\varepsilon))\psi_i, M_+(\varepsilon)> = <(\xi_1 - s(\varepsilon))\psi_i, M_-> \ . \tag{2.10}$$

in which $<F,G> = \int FGd\xi$, and $\psi_i = 1,\ \xi_i$ or ξ^2. Thus

$$\frac{d}{d\varepsilon} <(\xi_1 - s(\varepsilon))\psi_i,\ M_+(\varepsilon) - M_-> = 0 \ . \tag{2.11}$$

For $\varepsilon = 0$ this yields

$$<(\xi_1 - s_0)\psi_i,\ \Psi_0> = 0 \ . \tag{2.12}$$

Moreover since $\Psi_0 = \frac{d}{d\varepsilon} M_+(\varepsilon) = \frac{d\rho}{d\varepsilon}\frac{dM_+}{d\rho} + \frac{d\underset{\sim}{u}}{d\varepsilon}\frac{dM_+}{d\underset{\sim}{u}} + \frac{dT}{d\varepsilon}\frac{dM_+}{dT}$ (the parameters ρ, $\underset{\sim}{u}$, T represent the only possible variations of M) then

$$\Psi_0 = (b_0 + b_i\xi_i + b_4\xi^2)M_- = b_i\psi_i M_- \tag{2.13}$$

for some constants b_i. Substitute (2.13) into (2.12) to obtain five homogeneous equations for the five unknowns b_i with a parameter s_0:

$$\sum_{j=0}^{4} \langle (\xi_1 - s_0)\psi_i \, , \, \psi_j M \rangle \, b_j = 0 \qquad i=0,\cdots,4 \tag{2.14}$$

with $\{\psi_0 = 1, \psi_i = \xi_i, \psi_4 = \xi^2\}$. Explicitly (for $\rho_- = T_- = 1$, $u_- = 0$)

$$\begin{bmatrix} -s_0 & 1 & 0 & 0 & -3s_0 \\ 1 & -s_0 & 0 & 0 & 5 \\ 0 & 0 & -s_0 & 0 & 0 \\ 0 & 0 & 0 & -s_0 & 0 \\ -3s_0 & 5 & 0 & 0 & 15s_0 \end{bmatrix} \begin{bmatrix} b_0 \\ b_1 \\ b_2 \\ b_3 \\ b_4 \end{bmatrix} = 0 \tag{2.15}$$

This implies $b_2 = b_3 = 0$ and

$$\begin{bmatrix} -s_0 & 1 & -3s_0 \\ 1 & -s_0 & 5 \\ 3s_0 & 5 & -15s_0 \end{bmatrix} \begin{bmatrix} b_0 \\ b_1 \\ b_4 \end{bmatrix} = 0 \tag{2.16}$$

Thus

$$0 = \begin{vmatrix} -s_0 & 1 & -3s_0 \\ 1 & -s_0 & 5 \\ -3s_0 & 5 & -15s_0 \end{vmatrix}$$

$$= -2s_0(3s_0^2 - 5)$$

which implies that $s_0 = 0$ or $s_0 = \pm\sqrt{5/3}$. The second and third roots turn out to be most interesting. For arbitrary u_-, T_- and ρ_- they are $s_0 = u_- \pm(\frac{5}{3} T_-)^{1/2}$, which is called the sound speed. The corresponding Ψ_0 is a sound mode.

With s_0 and Ψ_0 chosen as above, we continue the description of $F = F_- + \varepsilon f(y = \varepsilon x, \xi)$. Proceed by a trick. By adding a correction Ψ_1 to Ψ_0, one can obtain

$$\Psi_\varepsilon = \Psi_0 + \varepsilon\Psi_1 \tag{2.17}$$

satisfying

$$L\Psi_\varepsilon = -\varepsilon\tau(\xi_1 - s)\Psi_\varepsilon + O(\varepsilon^2) \tag{2.18}$$

$$\langle\Psi_\varepsilon, (\xi_1 - s)\Psi_\varepsilon\rangle = \int M^{-1}(\xi_1 - s)\Psi_\varepsilon^2 d\xi = -\varepsilon\ . \tag{2.19}$$

The scaling in the normalization is correct since $\langle\Psi_0, (\xi_1 - s)\Psi_0\rangle = 0$. This is an (approximate) generalized eigenvalue problem; the solution Ψ_ε and τ is determined uniquely.

Now expand f in the form

$$f(y,\xi) = z(y)\Psi_\varepsilon(\xi) + \varepsilon w(y,\xi) \tag{2.20}$$

in which w satisfies

$$\langle(\xi_1 - s)\Psi_\varepsilon, w\rangle = 0 . \tag{2.21}$$

It follows that

$$\begin{aligned}\langle\Psi_\varepsilon, Lw\rangle &= \langle L\Psi_\varepsilon, w\rangle \\ &= \langle\varepsilon\tau(\xi_1 - s)\Psi_\varepsilon, w\rangle + O(\varepsilon^2) \\ &= O(\varepsilon^2)\end{aligned} \tag{2.22}$$

and

$$\begin{aligned}\langle\Psi_\varepsilon, Q(\Psi_\varepsilon, w)\rangle &= \langle\Psi_0, Q(\Psi_\varepsilon, w)\rangle + O(\varepsilon) \\ &= O(\varepsilon) .\end{aligned} \tag{2.23}$$

To find an equation for z take the inner product of the Boltzmann equation with Ψ_ε, i.e.

$$\frac{\partial}{\partial y}\langle\Psi_\varepsilon, (\xi_1 - s)f\rangle = \frac{1}{\varepsilon}\langle\Psi_\varepsilon, Lf\rangle + \langle\Psi_\varepsilon, Q(f,f)\rangle . \tag{2.24}$$

Because of (2.17) - (2.23),

$$\langle \Psi_\varepsilon,(\xi_1 - s)f\rangle = \langle \Psi_\varepsilon,(\xi_1 - s)z\Psi_\varepsilon\rangle$$

$$= -\varepsilon z \tag{2.25}$$

$$\langle \Psi_\varepsilon,Lf\rangle = \langle \Psi_\varepsilon,-\varepsilon\tau(\xi_1 - s)z\Psi_\varepsilon\rangle + O(\varepsilon^3)$$

$$= \varepsilon^2\tau z \tag{2.26}$$

$$\langle \Psi_\varepsilon,Q(f,f)\rangle = \langle \Psi_\varepsilon,Q(\Psi_\varepsilon,\Psi_\varepsilon)\rangle z^2 + O(\varepsilon^2)$$

$$= -\varepsilon\gamma z^2 + O(\varepsilon^2) \quad \text{(definition of } \gamma) \ . \tag{2.27}$$

Group these together to get

$$\frac{\partial}{\partial y}z = -\tau z + \gamma z^2 + O(\varepsilon) \ . \tag{2.28}$$

If the error term $O(\varepsilon)$ is ignored, the solution is

$$z(y) = \frac{1}{2}\left(\frac{\tau}{\gamma}\right)\{\tanh(-\frac{1}{2}\tau y) + 1\} \tag{2.29}$$

and therefore the shock profile looks like

$$F(x,\xi) = M_- + \varepsilon\frac{1}{2}\left(\frac{\tau}{\gamma}\right)\{\tanh(-\frac{1}{2}\tau\varepsilon x) + 1\}\Psi_\varepsilon(\xi) \ . \tag{2.30}$$

This hyperbolic tangent solution is also the approximate form of a weak shock solution for the Navier-Stokes equations (1.30) - (1.32)

The following theorem [1] states existence of solutions of the Boltzmann shock profile equations (2.1), (2.2) for weak shocks and the validity of the Navier-Stokes equations for such weak shocks. Its proof is based on the leading order calculations presented above.

Theorem 2.1 *Let* $\rho_\pm$, $T_\pm$, $u_\pm$ *satisfy* (2.4), (2.6) *with* $|\rho_+ - \rho_-| + |T_+ - T_-| + |u_+ - u_-| = \varepsilon$ *sufficiently small. Let* $\rho_{NS}(x)$, $T_{NS}(x)$, $u_{NS}(x)$ *be a solution of the steady Navier-Stokes equations with values* $\rho_\pm$, $T_\pm$, $u_\pm$ *at* $x = \pm\infty$, *and let* $M_{NS} = M(\rho_{NS}, T_{NS}, u_{NS})$. *Then there is a solution F of the Boltzmann shock profile equations* (2.1), (2.2). *Moreover it satisfies*

$$||M_{NS} - F|| \leq c\varepsilon^2 \ . \tag{2.31}$$

Note 1) The norm in (2.31) is

$$||F|| = \sup_{\underset{\sim}{x}} \sup_{\underset{\sim}{\xi}} (1 + |\underset{\sim}{\xi}|)^r |F(x,\underset{\sim}{\xi})| \tag{2.32}$$

for any $r > 0$.

2) The order ε^2 error shows significant agreement between M_{NS} and F.

2.3 References

[1] R.E. Caflisch and B. Nicolaenko, Shock profile solutions of the Boltzmann equation, Comm. Math. Phys., 86 (1982) 161-194.

[2] R.E. Caflisch, The half-space problem for the Boltzmann equation at zero temperature. Comm. Pure Appl. Math., 38 (1985) 529-547.

[3] P.D. Lohn, and T.S. Lundgren. Strong shock structure, Phys. Fluids, 17 (1974), 1808-1815.

[4] H.M. Mott-Smith, The solution of the Boltzmann equation for a shock wave, Phys. Rev., 82 (1951), 885-892.

[5] I. Tamm, Width of high-intensity shock waves, Proc. (Trudy) Lebedev Phys. Inst., 29 (1965), 231-241.

Lecture 3 Boundary Layers

Near a spatial boundary the molecular distribution is expected to vary rapidly in the direction perpendicular to the boundary. If the variation in time and in parallel spatial directions is slow, the corresponding derivatives can be neglected and the spatial domain can be taken to be a half space. Also it is assumed that F is nearly a uniform Maxwellian M as in (1.2) so that nonlinear terms in $\tilde{F}$ can be neglected (to leading order). The resulting linearized Boltzmann equation (with the tilda dropped) is

$$\xi_1 \frac{\partial}{\partial x} F = LF \qquad x > 0 \tag{3.1}$$

As a boundary condition at x = 0, we specify the incoming particles ($\xi_1 > 0$), i

$$F(\xi) = G(\xi) \qquad x = 0 , \quad \xi_1 > 0 \tag{3.2}$$

with G prescribed. We are also able to prescribe the mass flux m, i.e.

$$\int \xi_1 \, F \, d\xi = m \, . \tag{3.3}$$

As shown below, conservation properties imply that m is independent of x. Finally as a boundary condition at $x = \infty$, we ask that F remain bounded or more specifically that

$$F \in D = \{F: (1+|\xi|)^{1/2} M^{-1/2} F \in L^\infty(\mathbb{R}_x^+, L^2(\mathbb{R}_\xi^3)) \, ,$$
$$M^{-1/2} F_x \in L^2_{\ell oc}(\mathbb{R}_x^+, L^2(\mathbb{R}_\xi^3))\} \, .$$

The Milne problem of kinetic theory is to solve (3.1) - (3.3) with specified distribution G and mass flux m. It will be assumed that G satisfies

$$\int_{\xi_1 > 0} (1 + |\xi|) M^{-1} G^2 \, d\xi = \kappa_G < \infty \, . \tag{3.4}$$

Our main results are the following:

Theorem 3.1 (Existence). For any $m \in \mathbb{R}$ and any G satisfying (3.4), there is a solution $F \in D$ for the Milne problem (3.1) - (3.3).

Theorem 3.2 (Uniqueness). For any given $m \in \mathbb{R}$ and a given G satisfying (3.4), there is only one solution of (3.1) - (3.3) in D.

Theorem 3.3 (Asymptotic Properties). Suppose that $F \in D$ satisfies (3.1) - (3.3) with $m \in G$ satisfying (3.4). Then

$$\lim_{x \to \infty} F = (a_\infty + m\xi_1 + b_{2\infty}\xi_2 + b_{3\infty}\xi_3 + c_\infty \xi^2) M \equiv F_\infty \tag{3.5}$$

exists, in the sense that $F - F_\infty$ is in $L^2(x,\xi)$.

These results were proved for the hard spheres gas by Maslova [4] and Bardos, Caflisch and Nicolaenko [1]. The results were extended to other intermolecular force laws by Cercignani [2]. For hard spheres the limit (3.5) is approached pointwise at an exponential rate in x_j; for softer force laws no rate has been established but it is expected to be an exponential of a power.

The proof of these theorems may be described as follows: First, existence for (3.1) - (3.3) is established on a finite interval $[0,\ell]$. Second, a priori estimates are proved for the finite-interval solution. Third, the limit $\ell \to \infty$ is taken to obtain a solution of (3.1) - (3.3) on $[0,\infty)$. The a priori bounds yield the asymptotic results of Theorem 3.3. Fourth, uniqueness is proved using a variant of the a priori bounds. In this lecture we shall only indicate several key steps in obtaining a priori bounds for the infinite interval. The exposition will follow that of Cercignani [2].

We begin by exhibiting three special solutions of (3.1) which combine a linear term and a constant term in x. Note that $\xi_2 M$, $\xi_3 M$, $(\xi^2 - 5T)M \in Rg(L)$ (in which T is the temperature in M). The three functins $L^{-1}(\xi_2 M)$, $L^{-1}(\xi_3 M)$, $L^{-1}(\xi^2 - 5T)$ appear in the Chapman-Enskog expansion and are important in the subsequent analysis of this lecture. Three exact solutions of (3.1) are

$$F_2 = x\xi_2 M + L^{-1}(\xi_2 M)$$

$$F_3 = x\xi_3 M + L^{-1}(\xi_3 M) \qquad (3.6)$$

$$F_4 = x(\xi^2 - 5T)M + L^{-1}(\xi^2 - 5T)M \ .$$

These represent boundary layer solutions with constant gradients in the transverse velocities and in temperature.

Now consider a solution F of (3.1) - (3.3) that is bounded at $x = \infty$. Integration of (3.1) over ξ shows that

$$\frac{d}{dx}\int \xi_1 \, F d\xi = 0 \ , \qquad (3.7)$$

i.e. that the solution has a constant mass flux M. On the other hand $\xi_1 M$ is an exact solution of (3.1). Thus $F - m(\rho T)^{-1}\xi_1 M$ is a solution of (3.1) with zero mass flux. In this way attention may be restricted to problems with $m = 0$, i.e.

$$\int \xi_1 \, F d\xi = 0 \ . \qquad (3.3')$$

The solution F of (3.1), (3.2), (3.3') may be decomposed as $F = w + q$ with $w \in Rg(L)$, $q \in Nu\ell(L)$ and

$$q(x,\xi) = (a(x) + b_2(x)\xi_2 + b_3(x)\xi_3 + c(x)\xi^2)M(\xi)$$

since $m = 0$. From this form of q it follows that

$$\int \xi_1 q^2 M^{-1} d\xi = 0 \tag{3.8}$$

$$\int \xi_1 \xi_2 q d\xi = \int \xi_1 \xi_3 q d\xi = \int \xi_1 (\xi^2 - 5T) q d\xi = 0 . \tag{3.9}$$

Several properties of w are found by integrating (3.1) multiplied by ξ_2, ξ_3, $(\xi^2 - 5T)$. Since these functions are orthogonal to q and L, it follows that

$$\int \xi_1 \xi_2 w d\xi = c_2$$

$$\int \xi_1 \xi_3 w d\xi = c_3 \tag{3.10}$$

$$\int \xi_1 (\xi^2 - 5T) w d\xi = c_4$$

in which c_2, c_3, c_4 are constants.

These constants may be shown to be zero as follows: Multiply (3.1) by $M^{-1}(L^{-1}(\xi_1 \xi_2 M)$ to obtain

$$\frac{d}{dx}\int M^{-1}L^{-1}(\xi_1,\xi_2M)\xi_1Fd\xi = \int M^{-1}L^{-1}(\xi_1\xi_2M)LFd\xi$$

$$= \int M^{-1}(\xi_1\xi_2M)Fd\xi$$

$$= \int \xi_1\xi_2wd\xi$$

$$= c_2 \ . \qquad (3.11)$$

This calculation used the self-adjointness of L with respect to the weight M^{-1}, as well as (3.9) and (3.10). On the other hand, since F is bounded, the derivative on the left side of (3.11) can be constant only if the constant is zero. This shows that $c_2 = 0$ and similar calculations shows that $c_3 = c_4 = 0$.

The key relation for F and w is the following:

$$\int \xi_1F^2M^{-1}d\xi = \int \xi_1w^2M^{-1}d\xi \ . \qquad (3.12)$$

Just as $\int F^2M^{-1}d\xi$ is the linearization (around M) of the H-functional [3], the quantity $\int \xi_1F^2M^{1-}d\xi$ represents the linearized flux of H. Thus (3.12) is the statement that the (linearized) H-flux is all in the non-fluid-dynamic part of F. The relation (3.12) follows immediately from (3.8) and (3.10) with $c_2 = c_3 = c_4 = 0$.

Finally an indication of the usefulness of (3.12) is seen by considering a simple energy estimate for (3.1). Multiply (3.1) by $M^{-1}F$ to obtain

$$\frac{1}{2}\frac{d}{dx}\int \xi_1 F^2 M^{-1} d\xi = \int M^{-1} FLFd\xi \ . \tag{3.13}$$

Sinc L is self-adjoint (with weight M^{-1}) and LF = Lw

$$\int M^{-1}FLFd\xi = \int M^{-1}wLwd\xi$$

$$\leq -\bar{\nu}\int (1+|\xi|)^{\gamma} w^2 M^{-1} d\xi \ . \tag{3.14}$$

Using relation (3.12), the energy inequality becomes

$$\frac{1}{2}\frac{d}{dx}\int \xi_1 w^2 M^{-1} d\xi \leq -\bar{\nu}\int (1+|\xi|)^{\gamma} w^2 M^{1-} d\xi \ . \tag{3.15}$$

The utility of (3.12) is that it allows (3.15) to be written entirely in terms of w (without q). From (3.15) energy-like bounds and decay properties can be derived [1,2].

References

[1] C. Bardos, R. Caflisch and B. Nicolaenko. The Milne and Kramers Problems for the Boltzmann Equation of a Hard Sphere Gas, Comm. Pure and Appl. Math 39 (1986).

[2] C. Cercignani. Half Space Problems in the Kinetic Theory of Gases. in Proc. Symp. of Soc. for Interaction of Math. and Mech. (1985).

[3] F.A. Grunbaum, "Linearization for the Boltzmann Equation", Trans. AMS 165 (1972) 425-449.

[4] N.B. Maslova. "Kramers Problem in the Kinetic Theory of Gases" USSR Comp. Math. Phys. 22 (1982) 208-219.

Lecture 4 Discrete Velocity Models of the Boltzmann Equation

Discrete velocity models of the Boltzmann equation are important for several physical and mathematical reasons. First they provide simple examples for kinetic theory. Second they enjoy a well-developed theory of existence, uniqueness and asymptotic properties [4]. Third they may serve as aids in computation or analysis of the full Boltzmann equation. Finally discrete velocity models provide motivation and a method of analysis for cellular automata simulations of the Navier-Stokes equations [2].

In this lecture the framework is set for a theory of discrete velocity models that is new in several respects. The objective is to find a large set of such models, all of which have the usual conservation and equilibration properties, and to find the space of fluid dynamic equations corresponding to these models. The fluid dynamic equations are characterized by the equation of state and the viscosity and heat conductivity coefficients.

Previous constructions [1,3] have found only a few discrete velocity models. In the theory presented here many such models may be constructed because the summational invariants are required to be only

approximately equal to mass, momentum and energy.

The best known discrete velocity model is that of Broadwell [5].

One serious defect of this model, which is also true for other discrete

velocity models, is that the corresponding fluid equations have a

pressure which depends on velocity [6]. The main object of this

lecture is to construct a sequence of discrete velocity models with N

velocities, for which the model fluid dynamic equations converge to the

true Euler equations as $N \to \infty$.

4.1   Model Equations and Summational Invariants

Let $\xi = (\xi_1, \xi_2, \xi_3) \in V_N = \{\xi^{(1)}, \dots, \xi^{(N)}\}$, a set with N elements,

and let $\xi \in \mathbb{R}^3$, $t \in \mathbb{R}^+$. Denote $F = F(\underset{\sim}{x}, t, \xi)$ and $F_k(\underset{\sim}{x}, t) = F(\underset{\sim}{x}, t, \xi^{(k)})$.

The model equations are

$$\left(\frac{\partial}{\partial t} + \xi^{(k)} \cdot \frac{\partial}{\partial \underset{\sim}{x}}\right) F_k = Q_k(F,F) = \sum_{\ell mn} q_{k\ell mn}(F_m F_n - F_k F_\ell) . \qquad (4.1)$$

The coefficients q are assumed to have the following two properties

$$q_{k\ell mn} \geq 0 \qquad\qquad \text{(non-negativity)} \qquad (4.2)$$

$$q_{k\ell mn} = q_{\ell kmn} = q_{mnk\ell} . \qquad \text{(symmetry)} \qquad (4.3)$$

Next let $\psi^{(1)}, \dots, \psi^{(m)}$ be real valued functions on V_N, the summational invariants. We say that $((\xi^{(k)},\xi^{(\ell)}),\ (\xi^{(m)},\xi^{(n)}))$ are collisional pairs whenever

$$\psi^{(i)}(\xi^{(k)}) + \psi^{(i)}(\xi^{(\ell)}) = \psi^{(i)}(\xi^{(m)}) + \psi^{(i)}(\xi^{(n)}) \tag{4.4}$$

for $i=1,\cdots,p$. We may as well assume that $\psi^{(1)} \equiv 1$. It is assumed that

$$q_{k\ell mn} > 0 \tag{4.5}$$

if and only if $(\xi^{(k)},\xi^{(\ell)}),\ (\xi^{(m)},\xi^{(n)})$ are collisional pairs. Second it is assumed that $\{\psi^{(i)}\}$ is a complete system of summational invariants, by which we mean that any function g satisfying

$$g(\xi^{(k)}) + g(\xi^{(\ell)}) = g(\xi^{(m)}) + g(\xi^{(n)}) \tag{4.6}$$

for all collisional pairs must be a linear combination of the $\psi^{(i)}$ with

$$g(\xi) = \sum_{i=1}^{M} \alpha^{(i)}\psi^{(i)}(\xi) \qquad \forall\xi \in V_N \tag{4.7}$$

with $\alpha^{(i)}$ constants.

The assumptions for this class of models are non-negativity and symmetry of q, positivity of q only for collisional pairs, and completeness of the summational invariants. In general it is a difficult and unsolved technical problem to characterize the collisional pairs and to verify completeness.

4.2 Equilibria and conserved quantities

Let ϕ be any function on V_N. The symmetry (4.3) of q implies that

$$<\phi,Q(F,G)> \equiv \sum_k \phi_k Q_k(F,G)$$

$$= \sum_{k\ell mn} q_{k\ell mn}\, \phi_k (F_m G_n + F_n G_m - F_k G_\ell - F_\ell G_k)$$

$$= \frac{1}{4} \sum_{k\ell mn} q_{k\ell mn} (\phi_k + \phi_\ell - \phi_m - \phi_n) \tag{4.8}$$

$$\cdot\, (F_m G_n + F_n G_m - F_k G_\ell - F_\ell G_k)\ .$$

Several results follow from (4.8): First

$$<\psi^{(i)},Q(F,G)> = 0 \quad \text{for } i=1,\cdots,p\ . \tag{4.9}$$

Second

$$<\log F,Q(F,F)> = \frac{1}{4} \sum_{k\ell mn} q_{k\ell mn} \log(F_k F_\ell / F_m F_n) \cdot (F_m F_n - F_k F_\ell)$$

$$\leq 0 \tag{4.10}$$

with equality iff for every collisional pair

$$F_k F_\ell = F_m F_n\ . \tag{4.11}$$

Thus any equilibrium (model Maxwellian) state $\tilde{M}$ satisfying $Q(\tilde{M},\tilde{M}) = 0$ also satisfies (4.11). By taking the log of (4.11) and using the completeness of the collisional invariants, we see that all such $\tilde{M}$ are given by

$$\tilde{M}(\xi) = \exp\Big(\sum_{i=1}^{N} a_i \psi^{(i)}(\xi)\Big) . \tag{4.12}$$

Next (4.10) leads to the usual H-theorem, that if F is a solution of the spatially homogeneous equation (4.1) then

$$\frac{d}{dt} \sum_{k=1}^{N} F_k \log F_k \leq 0 \tag{4.13}$$

with equality iff $F = \tilde{M}$ satisfying (4.12).

The symmetry property (4.8) and equilibrium property (4.11) show that for any equilibrium state $\tilde{M}$ and any function ϕ

$$\langle\phi, Q(\phi\tilde{M},\tilde{M})\rangle = -\frac{1}{4} \sum_{k\ell mn} q_{k\ell mn}(\phi_k + \phi_\ell - \phi_m - \phi_n)^2 \tilde{M}_k \tilde{M}_\ell$$

$$\leq 0 \tag{4.14}$$

with equality iff $\phi = \psi^{(i)}$ (for some i), a summational invariant.

The macroscopic variables corresponding to a density function F are defined by

$$\rho^{(r)}(x,t) = \langle\psi^{(r)},F\rangle(x,t) = \sum_{k=1}^{N} \psi^{(r)}(\xi^{(k)}) F(\underline{x},t,\xi^{(k)}) . \tag{4.15}$$

An equilibrium distribution is determined by its macroscopic variables as stated in the following proposition.

Proposition 4.1. *If* $\tilde{M}^{(1)}$ *and* $\tilde{M}^{(2)}$ *are (positive) equilibrium distributions satisfying* (4.12) *and if*

$$\langle\psi^{(r)},\tilde{M}^{(1)}\rangle = \langle\psi^{(r)},\tilde{M}^{(2)}\rangle \tag{4.16}$$

for $r = 1,\cdots,p$, *then* $\tilde{M}^{(1)} \equiv \tilde{M}^{(2)}$.

Proof. Suppose that $\tilde{M}^{(1)} \neq \tilde{M}^{(2)}$. Denote $F(\alpha) = \alpha\tilde{M}^{(1)} + (1-\alpha)\tilde{M}^{(2)}$ and $H(F) = \sum_k F_k \log F_k$. Then $F'(\alpha)$ satisfies $F'(\alpha) \neq 0$ and $\langle\psi^{(r)},F'(\alpha)\rangle = 0$ $\forall\alpha$ and $F''(\alpha) = 0$. Thus

$$\begin{aligned}\frac{\partial}{\partial\alpha} H(F(0)) &= \sum_k (1 + \log F_k(0))F_k'(0) \\ &= \sum_k(\psi_k^{(1)} + \sum_{i=1}^{M} a_i^{(1)}\psi_k^{(i)})F_k'(0) \\ &= 0\end{aligned} \tag{4.17}$$

and similarly $\frac{\partial}{\partial\alpha} H(F(1)) = 0$. On the other hand

$$\begin{aligned}\frac{\partial^2}{\partial\alpha^2} H(F(\alpha)) &= \sum_k F_k(\alpha)^{-1}F_k'(\alpha)^2 \\ &> 0\end{aligned} \tag{4.18}$$

since $F_k(\alpha) > 0$ for $0 \leq \alpha \leq 1$. This contradiction concludes the proof.

4.3 Model Euler Equations

Since the collision operator Q is orthogonal to the collision invariants (cf.(4.9)), inner product of the model Boltzmann equation (4.1) with $\psi^{(r)}$ results in

$$\frac{\partial}{\partial\tau} \langle\psi^{(r)},F\rangle + \frac{\partial}{\partial \underset{\sim}{x}} \cdot \langle\xi\psi^{(r)},F\rangle = 0 . \qquad (4.19)$$

The first term in (4.19) is just the macroscopic variable $\rho^{(r)}$. In the small mean free path limit, F is nearly a (model) Maxwellian $\tilde{M} = \tilde{M}(\rho)$. Denote

$$\underset{\sim}{S}^{(r)}(\rho^{(1)},\cdots,\rho^{(m)}) = \langle\xi\psi^{(r)},M(\rho^{(1)},\cdots,\rho^{(m)})\rangle . \qquad (4.20)$$

Then in the small mean free path limit, (4.19) becomes

$$\frac{\partial}{\partial t} \rho^{(r)} + \frac{\partial}{\partial \underset{\sim}{x}} \cdot \underset{\sim}{S}^{(r)}(\rho^{(1)},\cdots,\rho^{(p)}) = 0 \qquad (4.21)$$

for $r = 1,\cdots,p$.

4.4 Example

The aim here is to construct a sequence of discrete velocity models that converge, at least in a formal sense, to the full Boltzmann equation. The model fluid dynamic equations will converge to the true

Euler equations. The main result will be determination of the convergence rate for the fluid equations.

In d dimensions let $\xi_{\underset{\sim}{r}} = n^{-1}\underset{\sim}{r}$ in which $\underset{\sim}{r} = (r_1,\cdots,r_d)$ with r_i integers and $-n^{1+\alpha} \le r_i \le n^{1+\alpha}$. This defines a velocity space containing $N = (2n^{1+\alpha} + 1)^d$ elements. For summational invariants define

$$\psi_0(\underset{\sim}{\xi}) = 1 \tag{4.22}$$

$$\psi_i(\underset{\sim}{\xi}) = \xi_i \qquad i = 1,\cdots,d \tag{4.23}$$

$$\begin{aligned}\psi_{d+1}(\underset{\sim}{\xi}) &= n^{-2+\beta}[n^{2-\beta}|\underset{\sim}{\xi}|^2] \\ &= n^{-2+\beta}[n^{-\beta}|\underset{\sim}{r}|^2] .\end{aligned} \tag{4.24}$$

In these definitions $\alpha > 0$ and $0 < \beta < 2$ are constants and $[\cdots]$ means "the integer part of."

The equilibria states $\tilde{M}$ are given, as in (4.12), by

$$\begin{aligned}\tilde{M}(\xi) &= \exp\{a + b_1\psi_1(\underset{\sim}{\xi}) + \cdots + b_d\psi_d(\underset{\sim}{\xi}) + c\psi_{d+1}(\underset{\sim}{\xi})\} \\ &= \rho(2\pi T)^{-3/2}\exp\{-\frac{1}{2T}(\psi_{d+1}(\xi) - \sum_1^d u_i\psi_i(\xi) + |\underset{\sim}{u}|^2)\}\end{aligned} \tag{4.25}$$

by a change of parameters. $\tilde{M}$ is written in (4.25) to look like an ordinary Maxwellian. In fact

$$|\psi_{d+1}(\underset{\sim}{\xi}) - |\underset{\sim}{\xi}|^2| = n^{-2+\beta}|[n^{2-\beta}|\underset{\sim}{\xi}|^2] - n^{2-\beta}|\underset{\sim}{\xi}|^2|$$

$$\leq n^{-2+\beta} . \tag{4.26}$$

It follows that

$$\tilde{M}(\xi) = M(\xi;\rho,\underset{\sim}{u},T)(1 + O(n^{-2+\beta})) . \tag{4.27}$$

Look at the model fluid dynamic equations. They are

$$\frac{\partial}{\partial t} \sum \tilde{M}_k + \frac{\partial}{\partial \underset{\sim}{x}} \cdot \sum \xi_k \tilde{M}_k = 0 \tag{4.28}$$

$$\frac{\partial}{\partial t} \sum \underset{\sim}{\xi}_k \tilde{M}_k + \frac{\partial}{\partial \underset{\sim}{x}} \cdot \sum \underset{\sim}{\xi}_k \xi_k M_k = 0 \tag{4.29}$$

$$\frac{\partial}{\partial t} \sum \psi_{d+1,k} \tilde{M}_k + \frac{\partial}{\partial \underset{\sim}{x}} \cdot \sum \xi_k \psi_{d+1,k} \tilde{M}_k = 0 . \tag{4.30}$$

Because of (4.27) these differ from the standard Euler equations (1.19) - (1.21) by an amount $O(n^{-2+\beta})$. Discretization errors, from replacing the sun by an integral, are smaller, i.e. $O(n^{-2})$.

An optimal choice of α and β should be made with the following considerations in mind. If there are too few collisional pairs then agreement of the model Boltzmann equation with the model Euler equations will be poor. On the other hand if there are too many collisional pairs, then some degrees of freedom have been wasted. In any case there are at least cn^d distinct velocities, i.e. $N \geq cn^d$ and the error e in the fluid equations is at least of size cn^{-2}.

We conclude that for this family of discrete velocity models with

N distinct velocities, the error in the fluid equations is no better than

$$e = cN^{-2/d} \tag{4.31}$$

in which d is the dimension and c is a constant. Although these results have been presented only formally, we conjecture that they are rigorously valid and that the error size $N^{-2/d}$ is optimal.

4.5 References

[1] H. Cabannes. The Discrete Boltzmann Equation (Theory and Applications) Lecture Notes at Univ. California, Berkeley 1980.

[2] U. Frisch, B. Hasslacher, and Y. Pomeau. Lattice Gas Automata for the Navier Stokes Equations. Phys. Rev. Letter 56 (1986) 1505.

[3] R. Gatignol. Theorie Critique des Gaz à Rèpartition Discrete de Vitesses, Lecture Notes in Physics #36 (1975) Spinger-Verlag.

[4] T. Platkowski and R. Illner. Discrete Velocity Models of the Boltzmann Equation: A Survey on the Mathematical Aspects of the Theory. SIAM Review to appear.

PHYSICO-CHEMICAL GAS DYNAMICS

J.F. Clarke
Cranfield Institute of Technology, Cranfield, Bedford, U.K.

1. INTRODUCTION

1.1 Boltzmann Equation for Polyatomic Molecules

When two particles, atoms or polyatomic molecules, collide there is always a possibility that one outcome of the event will be a redistribution of the energy contained in the internal structure of either or both collision partners. Such molecular encounters are called inelastic collisions to distinguish them from their less complicated elastic collision companions, in which no such redistributions of internal energy take place. Elastic collisions conserve all of the external manifestations of molecular presence, namely molecular mass, molecular momentum and molecular translational energy and, in addition, they conserve the identities of the two collision partners. Inelastic collisions also preserve mass and momentum but, where energy is concerned, it is the sum of molecular translational and internal energies that is preserved. Insofar as an inelastic collision between a pair of molecules may result in the emergence from the encounter of two molecules that are chemically totally different from the original incident pair, it is clear that inelastic encounters may not preserve identity. Indeed if one molecule is identified by both its chemical type and by its internal quantum state it can be said that identity is always lost in an inelastic collision, since internal quantum state must change when internal energy is exchanged in an encounter.

In order to make it clear in the developments that follow when a molecule is being referred to in terms of both its chemical type and by the set of numbers that go to define its internal quantum condition we shall refer to it as belonging to a particular 'species' $\underline{\alpha}$, $\underline{\beta}$, etc, as the case may be. The labels $\underline{\alpha}$, $\underline{\beta}$, etc, will always appear with an underline, and the word 'species' will always be written in inverted commas. In essence this notation is just a slightly more complicated version of the notation proposed for a description of polyatomic gas behaviour by Hirschfelder, et al, (1954) and others. It has the advantage for present purposes that it will make for easy reference to the true chemical type of molecule, regardless of its internal condition, by first describing it as belonging to a chemical species and second using α, β etc, as appropriate chemical species labels. Examples of the way in which this notation is to be used below should make the matter clear.

Furthermore, with Greek letters used to define both 'species' and chemical species, the Roman lower-case letters i, j, k etc can be used to indicate vector or tensor components in the usual manner of the Cartesian tensor system (Jeffreys, 1952).

In referring to the internal quantum state of molecules one is evidently recognizing the need to describe microscopic molecular behaviour by the rules of quantum mechanics rather than by those of classical Newtonian or Hamiltonian mechanics. According to Bohr's Correspondence Principle the distinction between classical and quantal behaviour disappears when the spacings between adjacent (quantised) energy levels is small compared with a datum energy given by the product kT of Boltzmann's constant k ($1.38 \times 10^{-23} JK^{-1}$) and absolute temperature T (in degrees Kelvin, K). For the translational motion of a hydrogen atom, essentially the worst case, in a box of one centimetre cube at ordinary temperatures this ratio is of order 10^{-16}(Clarke & McChesney 1964, §3.9) and it is evidently perfectly justifiable to treat the translational motion of molecules as a classical continuously distributed mode of energy storage.

This last fact makes it both reasonable and sensible to adopt a hybrid model of molecular behaviour, in which translational motion behaves classically whilst internal (microscopic) molecular structure follows the proper dictates of quantum theory. Since there is a fixed, constant, amount of energy per unit mass $e_{\underline{\alpha}}^{int}$ associated with 'species' $\underline{\alpha}$ in such a case it follows that one can, in essence, simply take over the Boltzmann equation for the one-particle molecular velocity distribution function $f_{\underline{\alpha}}$ directly, and use it to describe 'species' behaviour. At least, this statement is true provided that one is prepared to make an important generalisation that concerns the collision integrals. In any molecular encounter, of whatever kind and number of participating molecules, a <u>total</u> number of particles will be lost in a time interval dt from the phase-space 'volume' $dx_1dx_2dx_3$, $dv_1dv_2dv_3$, or dxdv for brevity, given by an expression $C_{\underline{\alpha}}^{(-)}dxdvdt$; similarly a total number $C_{\underline{\alpha}}^{(+)}dxdvdt$ may be gained in the same way, where x_i denotes a space vector and v_j is a molecular-velocity vector. There is no need to restrict the meaning of the collisional gain or loss terms $C_{\underline{\alpha}}^{(+)}$ or $C_{\underline{\alpha}}^{(-)}$ to binary collisions; encounters of any degree of complexity are implied by the use of these symbols. As a consequence one can write down a form of Boltzmann's equation that models behaviour of the one-particle distribution function $f_{\underline{\alpha}}$ in the context of the proposed hybrid classical/quantal description, without restriction as to the meaning of the collision terms.

A starting point for the analysis of a general reacting mixture of gases is therefore given by the relation

$$\frac{\partial f_{\underline{\alpha}}}{\partial t} = -v_{\underline{\alpha}k}\frac{\partial f_{\underline{\alpha}}}{\partial v_k} - F_{\underline{\alpha}k}\frac{\partial f_{\underline{\alpha}}}{\partial v_k} + [C_{\underline{\alpha}}^{(+)} - C_{\underline{\alpha}}^{(-)}], \qquad (1.1.1)$$

where $F_{\underline{\alpha}}$ is the external force acting on a particle of 'species' $\underline{\alpha}$. The fact that one is content with a one-particle distribution function implies that knowledge of the relative positions of two or more molecules is not important and that, as a consequence, one must be dealing only with so-called <u>dilute</u> gas mixtures.

1.2 Properties of Dilute Mixtures Based on $f_{\underline{\alpha}}$

The various quantities of macroscopic importance such as density, flow velocity, energy flux and so on, can be defined in terms of $f_{\underline{\alpha}}$ in the usual way. They are listed here for future reference, with the number density of 'species' $\underline{\alpha}$ written as $n_{\underline{\alpha}}$ and the mass of an $\underline{\alpha}$-'species' molecule written as $m_{\underline{\alpha}}$.

As one might expect with a mixture of gases, consisting of many different 'species' $\underline{\alpha}$, some new phenomena appear that are not present in simple single monatomic gases. Some of the most important of these are the appearance of diffusion of 'species' through the mixture, the need to deal separately with energies internal to the molecule and the energy contained in translation and, as will be seen in §1.3, the need for a conservation equation for each 'species' in the mixture. Clearly the average value of any property $\varphi_{\underline{\alpha}}$, say, of 'species' $\underline{\alpha}$ at some point in physical space and time will be given by $\bar{\varphi}_{\underline{\alpha}}$ where $\bar{\varphi}_{\underline{\alpha}} n_{\underline{\alpha}} = \int \varphi_{\underline{\alpha}} f_{\underline{\alpha}} dv_{\underline{\alpha}}$.

Average velocity of $\underline{\alpha}$; $\qquad \bar{v}_{\underline{\alpha}j} = \frac{1}{n_{\underline{\alpha}}} \int v_{\underline{\alpha}j} f_{\underline{\alpha}} dv_{\underline{\alpha}}$; (1.2.1)

Density of mixture; $\qquad \rho = \sum_{\underline{\alpha}} m_{\underline{\alpha}} n_{\underline{\alpha}} = \sum_{\underline{\alpha}} \rho_{\underline{\alpha}}$; (1.2.2)

Mass-average (flow) velocity; $\qquad u_j = \frac{1}{\rho} \sum_{\underline{\alpha}} \rho_{\underline{\alpha}} \bar{v}_{\underline{\alpha}j}$; (1.2.3)

Peculiar velocity; $\qquad V_{\underline{\alpha}j} = v_{\underline{\alpha}j} - u_j$; (1.2.4)

Diffusion (mean peculiar) velocity of $\underline{\alpha}$; $\qquad u_{\underline{\alpha}j} = \bar{v}_{\underline{\alpha}j} - u_j$; (1.2.5)

Translational energy per unit mass of $\underline{\alpha}$;

$$= \frac{1}{\rho} \int (\tfrac{1}{2} m_{\underline{\alpha}} V_{\underline{\alpha}k} V_{\underline{\alpha}k}) f_{\underline{\alpha}} dV_{\underline{\alpha}} = \tfrac{1}{2}\overline{V_{\underline{\alpha}k} V_{\underline{\alpha}k}} = e^{tr}_{\underline{\alpha}} \; ; \qquad (1.2.6)$$

(NB the use of the Cartesian tensor summation convention; a repeated subscript as in $u_k u_k$ for example implies summation over all three values of k, i.e. $u_k u_k \equiv u_1^2 + u_2^2 + u_3^2$.)

Translation energy per unit mass of mixture,

$$e^{tr} = \frac{1}{\rho} \sum_{\underline{\alpha}} \rho_{\underline{\alpha}} e_{\underline{\alpha}}^{tr} . \tag{1.2.7}$$

If n, where

$$n = \sum_{\underline{\alpha}} n_{\underline{\alpha}} , \tag{1.2.8}$$

is the total number of molecules in unit volume of the mixture the <u>kinetic theory translational temperature</u> T_1 is defined so that

$$\frac{3}{2} nkT_1 \equiv \sum_{\underline{\alpha}} \rho_{\underline{\alpha}} e_{\underline{\alpha}}^{tr} = \rho e^{tr} \tag{1.2.9}$$

where k is Boltzmann's constant.

Intrinsic* energy per unit mass of mixture,

$$e = e^{tr} + \frac{1}{\rho} \sum_{\underline{\alpha}} m_{\underline{\alpha}} n_{\underline{\alpha}} e_{\underline{\alpha}}^{int} \tag{1.2.10}$$

In using (1.2.10) it is implied that there is no (significant) potential energy in the system that is attributable to the relative <u>positions</u> of the molecules. This is wholly in keeping with the idea that the gas mixture is dilute (cf §1.1), and the relative unimportance of potential energy can indeed be used as a definition of the term dilute.

*Footnote. The usual word here is "internal"; however, when one is forced to distinguish between energy that is present on account of the internal structure of a molecule and energy that is present on account only of translational motion of the molecule, it is very useful to be able to describe the energy of the whole molecule as "intrinsic" and the energy of the internal structure as "internal".

The molecules in such a mixture exist independently of their companions and, as a consequence, their contributions to the mixture's energy and momentum are simply additive. This fact is exploited in the treatment of fluxes, as will now be demonstrated.

Suppose that $\psi_{\underline{\alpha}}$ denotes a scalar property of a molecule, such as its mass, its energy or one of the three components of its momentum vector, and consider the rate of transport of this property at a given instant of time and a given point in space across an element of area dA moving with the flow velocity u_j, and oriented in the direction of a unit vector $\tilde{n}_j$. The rate at which the property is carried across dA at a given instant is given by $\psi_{\underline{\alpha}}(f_{\underline{\alpha}} dV_{\underline{\alpha}})\tilde{n}_k V_{\underline{\alpha}k}\, dA$, where $(f_{\underline{\alpha}} dV_{\underline{\alpha}})$ is the number of molecules of $\underline{\alpha}$ per unit volume of physical space with peculiar velocities in the range $dV_{\underline{\alpha}}$ ($\equiv dV_{\underline{\alpha}1}dV_{\underline{\alpha}2}dV_{\underline{\alpha}3}$) about $V_{\underline{\alpha}j}$ and $\tilde{n}_k V_{\underline{\alpha}k}$ is the component of $V_{\underline{\alpha}j}$ perpendicular to dA. The instantaneous total rate of transport of the property per unit area relative to the flow velocity u_j is therefore given by

$$\tilde{n}_k \left(\int \psi_{\underline{\alpha}} V_{\underline{\alpha}k}\, f_{\underline{\alpha}} dV_{\underline{\alpha}}\right) \equiv \tilde{n}_k (\Psi_{\underline{\alpha}k}) \qquad (1.2.11)$$

and it should be noted that $dV_{\alpha 1}$ is synonymous with $dv_{\alpha 1}$ etc, (see (1.2.4)) since u_j is fixed in present circumstances (i.e. t & x_j fixed). The quantity $\Psi_{\underline{\alpha}j}$ is a flux vector.

When $\psi_{\underline{\alpha}}$ is $m_{\underline{\alpha}}$, $\Psi_{\underline{\alpha}j}$ is the diffusive mass flux vector $g_{\underline{\alpha}j}$ where

$$g_{\underline{\alpha}j} = m_{\underline{\alpha}} n_{\underline{\alpha}} u_{\underline{\alpha}j}. \qquad (1.2.12)$$

When $\psi_{\underline{\alpha}}$ is $m_{\underline{\alpha}}(\frac{1}{2}V_{\underline{\alpha}k} V_{\underline{\alpha}k} + e_{\underline{\alpha}}^{int})$, $\Psi_{\underline{\alpha}j}$ is the energy-flux vector for 'species' $\underline{\alpha}$, namely $q_{\underline{\alpha}j}$, where

$$q_{\underline{\alpha}j} = \tfrac{1}{2}m_{\underline{\alpha}}n_{\underline{\alpha}}\overline{V_{\underline{\alpha}k}V_{\underline{\alpha}k}V_{\underline{\alpha}j}} + m_{\underline{\alpha}}n_{\underline{\alpha}}u_{\underline{\alpha}j}e_{\underline{\alpha}}^{int} , \qquad (1.2.13)$$

since $e_{\underline{\alpha}}^{int}$ is constant for any given $\underline{\alpha}$.

Note that the internal energy of a molecule $e_{\underline{\alpha}}^{int}$ is transported at the diffusion velocity $u_{\underline{\alpha}j}$ in (vectorial) addition to its convection at the flow velocity u_j.

The total energy-flux vector is

$$q_i = \sum_{\underline{\alpha}} q_{\underline{\alpha}i} = \sum_{\underline{\alpha}}\{\tfrac{1}{2}m_{\underline{\alpha}}n_{\underline{\alpha}}\overline{V_{\underline{\alpha}k}V_{\underline{\alpha}k}V_{\underline{\alpha}i}} + m_{\underline{\alpha}}n_{\underline{\alpha}}u_{\underline{\alpha}i}e_{\underline{\alpha}}^{int}\} . \qquad (1.2.14)$$

If $\psi_{\underline{\alpha}}$ is one of the components of momentum $m_{\underline{\alpha}}V_{\underline{\alpha}1}$, $m_{\underline{\alpha}}V_{\underline{\alpha}2}$ or $m_{\underline{\alpha}}V_{\underline{\alpha}3}$ (1.2.11) can be used to find an element $p_{\underline{\alpha}1j}$, $p_{\underline{\alpha}2j}$ or $p_{\underline{\alpha}3j}$ of the partial-pressure tensor $p_{\underline{\alpha}ij}$ for 'species' $\underline{\alpha}$. In general it can be seen that

$$p_{\underline{\alpha}ij} = m_{\underline{\alpha}}n_{\underline{\alpha}}\overline{V_{\underline{\alpha}i}V_{\underline{\alpha}j}} . \qquad (1.2.15)$$

The partial-pressure tensor is symmetric, and the total pressure tensor is evidently given (cf the statement that follows (1.2.10) above) by

$$p_{ij} = \sum_{\underline{\alpha}} p_{\underline{\alpha}ij} , \qquad (1.2.16)$$

which is simply an extension of Dalton's Law of Partial Pressures. From the definition of $e_{\underline{\alpha}}^{tr}$ in (1.2.6) it can be seen that

$$p_{\underline{\alpha}11} + p_{\underline{\alpha}22} + p_{\underline{\alpha}33} = 2\rho_{\underline{\alpha}}e_{\underline{\alpha}}^{tr} \equiv 3p_{\underline{\alpha}} . \qquad (1.2.17)$$

Evidently $p_{\underline{\alpha}}$ defines the scalar thermodynamic partial pressure of 'species' $\underline{\alpha}$. In view of (1.2.9) it can be seen that

$$p = \sum_{\underline{\alpha}} p_{\underline{\alpha}} = \frac{2}{3}\sum_{\underline{\alpha}} \rho_{\underline{\alpha}}e_{\underline{\alpha}}^{tr} = \frac{2}{3}\rho e^{tr} = nkT_1 . \qquad (1.2.18)$$

This important result shows that the mixture's scalar thermodynamic pressure p is formally related to number-density and absolute temperature, in the present general nonequilibrium conditions, in exactly the same way as it is in conditions of complete equilibrium (see e.g. Vincenti & Kruger, 1965, Ch. 1 §3) provided that, under nonequilibrium conditions, the temperature T_1 refers to the energy-content of the translational modes of molecular motion alone.

It is appropriate to remark at this juncture that the concept of temperature should be treated with care in situations for which a lack of equilibrium, in the conventional thermodynamical and chemical sense, is axiomatic. It can be verified that T_1 defined in (1.2.9) is identical with the thermodynamic absolute temperature T under equilibrium conditions (Hirschfelder, et al, 1954, §7.2(a)), when all modes of molecular energy storage are described by T. The essence of non-equilibrium situations can be illustrated by remarking that the energy content of, say, the vibrational internal-molecular mode of motion of a molecule will usually not be known when just the translational temperature T_1 is known.

Suppose for the moment that one defines the translational temperature of 'species' $\underline{\alpha}$ as $T_{\underline{\alpha}1}$, where by analogy with (1.2.9)

$$\frac{3}{2}\, n_{\underline{\alpha}} k T_{\underline{\alpha}1} = \rho_{\underline{\alpha}} e_{\underline{\alpha}}^{tr} = \frac{3}{2}\, p_{\underline{\alpha}} \, , \qquad (1.2.19)$$

having used (1.2.17) to acquire the last result. It follows from (1.2.18) that

$$p = \sum_{\underline{\alpha}} p_{\underline{\alpha}} = \sum_{\underline{\alpha}} n_{\underline{\alpha}} k T_{\underline{\alpha}1} = k T_1 \sum_{\underline{\alpha}} n_{\underline{\alpha}} \qquad (1.2.20)$$

and that either $T_{\underline{\alpha}1}$ is equal to T_1, or $T_{\underline{\alpha}1}$ depends upon $n_{\underline{\alpha}}$ in some way; (1.2.19) shows that, if the latter is true,

$$e_{\underline{\alpha}}^{tr} = \frac{3}{2} \frac{k}{m_{\underline{\alpha}}} T_{\underline{\alpha}1} \tag{1.2.21}$$

will depend upon $n_{\underline{\alpha}}$ in some way. This is inconsistent with the independent character of the molecules in a dilute gas, and so

$$T_{\underline{\alpha}1} = T_1 \tag{1.2.22}$$

is the only acceptable possibility. Then (1.2.21) makes

$$e_{\underline{\alpha}}^{tr} = \frac{3}{2} \frac{k}{m_{\underline{\alpha}}} T_1 = \frac{3}{2} \frac{R}{W_{\underline{\alpha}}} T_1 \tag{1.2.23}$$

where R is the Universal Gas Constant and $W_{\underline{\alpha}}$ is the (dimensionless) molecular weight of $\underline{\alpha}$

Evidently (1.2.20) can now be written as

$$p = (\sum_{\underline{\alpha}} m_{\underline{\alpha}} n_{\underline{\alpha}} \frac{R}{W_{\underline{\alpha}}}) T_1 = \rho R T_1 \sum_{\alpha} (\frac{c_{\underline{\alpha}}}{W_{\underline{\alpha}}}) \tag{1.2.24}$$

where

$$c_{\underline{\alpha}} \equiv \rho_{\underline{\alpha}} / \rho \tag{1.2.25}$$

is the mass-fraction of 'species' $\underline{\alpha}$. The molecular weight of the mixture is given by W, where

$$\frac{1}{W} = \sum_{\alpha} (\frac{c_{\underline{\alpha}}}{W_{\underline{\alpha}}}) \tag{1.2.26}$$

and so (1.2.24) can be written simply as

$$p = \rho \frac{R}{W} T_1 . \tag{1.2.27}$$

Note that ρ can also be written in the form

$$\rho = \sum_{\underline{\alpha}} m_{\underline{\alpha}} n_{\underline{\alpha}} = \frac{1}{\tilde{N}} \sum_{\underline{\alpha}} W_{\underline{\alpha}} n_{\underline{\alpha}} \tag{1.2.28}$$

where $\tilde{N}$ is Avogadro's number. Then, since $\tilde{N}k$ is the same as R it follows that

$$p/\rho = (\sum_{\underline{\alpha}} n_{\alpha} / \sum_{\underline{\alpha}} n_{\underline{\alpha}} W_{\underline{\alpha}}) R T_1 = (n / \sum_{\underline{\alpha}} n_{\underline{\alpha}} W_{\underline{\alpha}}) R T_1 = R T_1 / \sum_{\underline{\alpha}} x_{\underline{\alpha}} W_{\underline{\alpha}} \tag{1.2.29}$$

where $x_{\underline{\alpha}}$ is $n_{\underline{\alpha}}/n$, the mole fraction of $\underline{\alpha}$. An alternative expression for mixture molecular weight is therefore

$$W = \sum_{\underline{\alpha}} x_{\underline{\alpha}} W_{\underline{\alpha}} \,. \qquad (1.2.30)$$

1.3 Equations of Change

One of the most important features of the Boltzmann equation (1.1.1) in the context of the flow of complicated gas mixtures, within which there are significant numbers of inelastic molecular collisions, is the capacity that it provides for a rigorous derivation of a set of conservation equations with which to describe macroscopic behaviour. All that is necessary is to multiply (1.1.1) by molecular mass, momentum or energy (sum of translational and internal) followed by integration of the whole equation over all translational molecular velocities. By this process of "taking moments" of the Boltzmann equation, and making use of the fluxes defined in (1.2.12, 14, 15 and 16), one can derive the following conservation equations.

'Species' Conservation:

$$\frac{\partial}{\partial t}\rho_{\underline{\alpha}} + \frac{\partial}{\partial x_k}(\rho_{\underline{\alpha}} u_k + g_{\underline{\alpha}k}) = m_{\underline{\alpha}}\dot{N}_{\underline{\alpha}} \equiv K_{\underline{\alpha}} \qquad (1.3.1)$$

where

$$\dot{N}_{\underline{\alpha}} \equiv \int [C_{\underline{\alpha}}^{(+)} - C_{\underline{\alpha}}^{(-)}]dv_{\underline{\alpha}} \qquad (1.3.2)$$

is the total rate of gain of 'species' $\underline{\alpha}$ molecules in unit volume in unit time, regardless of the individual molecular velocities. Evidently $K_{\underline{\alpha}}$ is the rate of increase of mass of $\underline{\alpha}$ per unit volume per unit time, and therefore represents the source-like effect of a generalized 'reaction' between molecules of all 'species'. One must remember that the [ ]-bracket terms in the integral (1.3.2) stand symbolically for all collisions between particles in the mixture that lead to the appearance of an $\underline{\alpha}$-'species' molecule in a differential volume of x_j, $v_{\underline{\alpha}j}$ phase-space; as such an expression like (1.3.2) is an exact expression for the 'reaction' rate.

Note that conservation of mass requires

$$\sum_{\alpha} m_{\alpha} \dot{N}_{\alpha} = \sum_{\alpha} K_{\alpha} = 0, \qquad (1.3.3)$$

whence (1.2.2) and (1.3.1) together provide the overall mixture mass-conservation equation

$$\frac{\partial}{\partial t}\rho + \frac{\partial}{\partial x_k}(\rho u_k) = 0, \qquad (1.3.4)$$

because

$$\sum_{\alpha} g_{\alpha j} = \sum_{\alpha} \rho_{\alpha} u_{\alpha j} = \sum_{\alpha} \rho_{\alpha} \overline{v}_{\alpha j} - u_j \sum_{\alpha} \rho_{\alpha} = 0 \qquad (1.3.5)$$

(see (1.2.2, 3 and 5)). In other words u_j describes the motion of the centre of mass of an element of the fluid and $u_{\alpha j}$ is then the velocity of 'species' α relative to this mass centre; equation (1.3.5) is expressive of this centre-of-mass concept.

It is very frequently useful to describe the composition of a mixture by specifying the mass-fractions c_{α} of all of the 'species' α where

$$\rho\, c_{\alpha} = \rho_{\alpha} . \qquad (1.3.6)$$

From (1.2.2) it is evident that

$$\sum_{\alpha} c_{\alpha} = 1 . \qquad (1.3.7)$$

Multiplying Boltzmann's equation by one or other of the three components of the linear momentum $m_{\alpha} v_{\alpha j}$ of a 'species' α molecule, integrating over all $v_{\alpha j}$, summing over all 'species' and using (1.2.2, 3, 15 and 16) gives the momentum equations

$$\frac{\partial}{\partial t}(\rho u_j) = -\frac{\partial}{\partial x_k}(p_{jk} + \rho u_j u_k) + \sum_{\alpha} \rho_{\alpha} F_{\alpha j}, \quad j = 1, 2, 3. \qquad (1.3.8)$$

Contributions from the collision terms vanish because momentum must be conserved in any kind of molecular encounter, be it elastic or inelastic in character. It is very frequently useful to recognize the role of the thermodynamic pressure p, defined in (1.2.18), in p_{ij} and so to write

$$p_{ij} = p\delta_{ij} - \tau_{ij} \tag{1.3.9}$$

where δ_{ij} is the unit tensor; τ_{ij} is called the viscous part of the stress tensor.

Multiplying (1.1.1) by the energy $\frac{1}{2}m_{\underline{\alpha}}v_{\underline{\alpha}k}v_{\underline{\alpha}k} + m_{\underline{\alpha}}e_{\underline{\alpha}}^{int}$ of a 'species' $\underline{\alpha}$ molecule, and proceeding as described in the previous paragraph with the aid of various definitions in §1.2 leads to the energy equation.

$$\frac{\partial}{\partial t}[\rho(e + \tfrac{1}{2}u^2)] = -\frac{\partial}{\partial x_k}\{\rho u_k(e + \tfrac{1}{2}u^2) + q_k + p_{jk}u_j\}$$
$$+ \sum_{\underline{\alpha}} \rho_{\underline{\alpha}}(u_k + u_{\underline{\alpha}k})F_{\underline{\alpha}k} \tag{1.3.10}$$

where intrinsic energy e is defined in (1.2.10), energy-flux vector q_j in (1.2.14) and, for brevity, $u_k u_k$ has been written simply as u^2.

Equations (1.3.1, 4, 8 and 10), which describe conservation of 'species', mass, momentum and energy, make up the Equations of Change in the title of this Section. They will enable us to predict the behaviour of dilute mixtures of general 'reacting' polyatomic gases once the various flux vectors $g_{\underline{\alpha}k}$, p_{ij} and q_j, and the 'reaction' source quantities $K_{\underline{\alpha}}$, have been related to the basic flow variables such as ρ, u_j, T_1 and p, for example. This can be done, at least in principle, by seeking a solution even if only implicitly, for the one-particle velocity distribution function $f_{\underline{\alpha}}$. The extraction of sufficient information to complete the kinetic theory formulation of macroscopic gas-dynamical processes with associated 'reactions' will be outlined in the next Chapter.

2. COLLISIONS, 'REACTIONS' AND FLUXES

2.1 Collision Integrals in the Boltzmann Equation

The influence of intermolecular collisions on the evolution of $f_{\underline{\alpha}}$ in six-dimensional phase space has been given only in symbolical form in (1.1.1). If one wishes to proceed further than mere derivation of equations of change, as described in §1.3, then completion of (1.1.1) by means of some explicit form for $C_{\underline{\alpha}}^{(+),(-)}$ is necessary. In particular, of course, it is clear that the set of flux vectors $g_{\underline{\alpha}j}$, q_j and the pressure tensor p_{ij}, defined in §1.2, as well as the generalised 'reaction'-rate of (1.3.2), all depend directly upon $f_{\underline{\alpha}}$ and hence upon the solution of (1.1.1). The essence of the treatment of polyatomic and reacting gases so far has required recognition of the distinguishing property of internal quantum state. Any description of the interactions between a pair of molecules, for example, must therefore take account of the possible transition from states $\underline{\alpha}$, $\underline{\beta}$ in the pre-collision pair to states $\underline{\gamma}, \underline{\delta}$ respectively in the post-collision particles. If the molecule in state $\underline{\alpha}$ remains in that state, <u>and</u> the molecule in state $\underline{\beta}$ remains in state $\underline{\beta}$ the encounter is an elastic one; otherwise it is inelastic.

If attention is fixed on an $\underline{\alpha}$ 'species' molecule a molecule of any other 'species' $\underline{\beta}$, say, will approach it with a relative velocity $g_{\underline{\alpha\beta}j}$ equal to $v_{\underline{\beta}j} - v_{\underline{\alpha}j}$. The number of $\underline{\beta}$ molecules in unit volume of real space around position x_j at time t with velocity in the range $dv_{\underline{\beta}}$ about $v_{\underline{\beta}}$ is given by $f_{\underline{\beta}}dv_{\underline{\beta}}$, and their rate of approach to the $\underline{\alpha}$-molecule will therefore be given by $f_{\underline{\beta}}dv_{\underline{\beta}}g_{\underline{\alpha\beta}}$ per unit area per unit time, where $g_{\underline{\alpha\beta}}$

is the modulus of $g_{\alpha\beta j}$. The ensuing interaction will cause the
β-'species' molecule to scatter into a solid angle $d\omega = \sin\Theta \, d\Theta \, d\phi$
centred about the direction Θ, ϕ (measured from the α-molecule as origin)
whilst it also undergoes a transition from state β to state δ; at the same
time α transits to γ. Molecule α presents a "target area" for β for this
process to take place; this quantity is written as $I_{\underline{\alpha\beta}}^{\gamma\delta}$ ($g_{\alpha\beta}$, Θ, ϕ) $\times$
$\sin\Theta \, d\Theta \, d\phi$ and $I_{\alpha\beta}^{\gamma\delta}$ is called the <u>differential</u> <u>cross-section</u> (or sometimes
just the cross-section) for Θ, ϕ scattering at relative speed $g_{\alpha\beta}$ with
α β to γ δ transition. The total rate at which this group of β-molecules
interacts with α-molecules in unit volume of real space is therefore

$$(f_{\underline{\beta}} dv_{\underline{\beta}} g_{\underline{\alpha\beta}})(I_{\underline{\alpha\beta}}^{\gamma\delta} \sin\Theta \, d\Theta \, d\phi) f_{\underline{\alpha}} dv_{\underline{\alpha}}$$

since $f_{\underline{\alpha}} dv_{\underline{\alpha}}$ is the number of α-molecules in unit volume of real space
with velocities in the range $dv_{\underline{\alpha}}$ about $v_{\underline{\alpha} j}$. Integrating over Θ from $-\frac{1}{2}\pi$
to $+\frac{1}{2}\pi$, over ϕ from 0 to 2π and over all three components of $v_{\underline{\beta} j}$,
followed by summation over all possible values of β, γ and δ evidently
gives the total rate at which α-molecules are lost per unit volume of
real space from 'volume' $dv_{\underline{\alpha}}$ about $v_{\underline{\alpha} j}$.

<u>If</u> one is prepared to accept that binary collisions alone are
significant, the foregoing will give the value of $c_{\underline{\alpha}}^{(-)} dv_{\underline{\alpha}}$ or, in other
words

$$c_{\underline{\alpha}}^{(-)} dv_{\underline{\alpha}} = \sum_{\underline{\beta}\underline{\gamma}\underline{\delta}} \int\int\int f_{\underline{\alpha}} f_{\underline{\beta}} g_{\underline{\alpha\beta}} I_{\underline{\alpha\beta}}^{\gamma\delta} \sin\Theta \, d\Theta \, d\phi \, dv_{\underline{\beta}} \, dv_{\underline{\alpha}}. \qquad (2.1.1)$$

By exactly analogous arguments the rate at which α-'species' molecules
accumulate in the range $dv_{\underline{\alpha}}$ about $v_{\underline{\alpha} j}$ can be written as

$$c_{\underline{\alpha}}^{(+)} dv_{\underline{\alpha}} = \sum_{\underline{\beta}\underline{\gamma}\underline{\delta}} \int\int\int f_{\underline{\gamma}}' f_{\underline{\delta}}' g_{\underline{\gamma}\underline{\delta}}' I_{\underline{\gamma}\underline{\delta}}'^{\alpha\beta} \sin\Theta \, d\Theta \, d\phi \, dv_{\underline{\gamma}} \, dv_{\underline{\delta}} \qquad (2.1.2)$$

where $v_{\underline{\gamma} j}'$, $v_{\underline{\delta} j}'$ are now pre-collision velocities that become $v_{\underline{\alpha} j}$, $v_{\underline{\beta} j}$
after being scattered by the collision whose cross-section is $I_{\underline{\gamma}\underline{\delta}}'^{\alpha\beta}$. The

$(\ )'$ denotes that the quantity depends upon the similarly-marked molecular velocities.

One must obviously relate $dv'_{\underline{\gamma}}\, dv'_{\underline{\delta}}$ to $dv_{\underline{\alpha}} dv_{\underline{\beta}}$ before completing one's estimate of the Boltzmann collision operator in (1.1.1), but (2.1.2) will suffice for now.

Such an estimate of the effect of collisions on the progress of $\underline{\alpha}$-'species' molecules through phase-space is subject to all of the familiar caveats, such as molecular chaos, negligible time-scale of interaction relative to mean time intervals between collisions, as well as the limitations to binary collisions and any implications that may arise in the matter of inverse-collisions. However the form just given will suffice for present purposes, especially since the complete physical situation is truly so complicated that one must make a "modelling" decision at some stage, and accept the implications that follow from such a decision. Some of these implications will be discussed in the work that follows.

2.2 The Infrequency of Inelastic Collisions

The generalized 'reaction' rate $\dot{N}_{\underline{\alpha}}$ defined in (1.3.2), will only receive a contribution from the integration of (2.1.1) and (2.1.2) over all molecular velocities of 'species' $\underline{\alpha}$ as a consequence of the presence of inelastic collisions ($\underline{\gamma} \neq \underline{\alpha}$ & $\underline{\delta} \neq \underline{\beta}$). It will therefore be important to know what fraction of all molecular collisions are in fact inelastic.

Some detailed comments about the latter will be made shortly and will indicate that, speaking in general terms, the rates of production of new 'species' associated with the 'reaction' sources $\dot{N}_{\underline{\alpha}}$ are mostly slow in terms of molecular collision frequencies. The very important inference that can be drawn from this conclusion is that when it is the 'reactive' collisions themselves that are the major influence in a flow the local $f_{\underline{\alpha}}$ functions will not be far removed from a local equilibrium value, based on local translational temperature T_1 (or, perhaps preferably in view of occasional ambiguities about the meaning of temperature, on local average translational energies; cf (1.2.23)). It then follows that many 'reactive' gas flows which involve significant sources of 'species' in the homogenous gas phase will be well described by the sort of collision-dominated, or continuum, theory that is the outcome of Chapman-Enskog solutions of Boltzmann's equation.

For present purposes this means that the local equilibrium, Maxwellian, distribution of molecular velocities will play a central role. Writing the one-particle distribution function $f_{\underline{\alpha}}$ as $f_{\underline{\alpha}}^{(o)}$ under these circumstances, it follows that

$$f_{\underline{\alpha}}^{(o)} = n_{\underline{\alpha}}\left(\frac{m_\alpha}{2\pi kT_1}\right)^{\frac{3}{2}} \exp\left\{-m_{\underline{\alpha}} V_{\underline{\alpha}k} V_{\underline{\alpha}k}/2kT_1\right\} \tag{2.2.1}$$

will form a basis both for Chapman-Enskog derivations of flux vectors and, with $n_{\underline{\alpha}}$ quite specifically not equal to an equilibrium value, for the calculation of $\dot{N}_{\underline{\alpha}}$ source terms. Note that it is useful to define a normalised local-equilibrium Maxwellian function $\bar{f}_{\underline{\alpha}}^{(o)}$ such that

$$f_{\underline{\alpha}}^{(o)} = n_{\underline{\alpha}}\bar{f}_{\underline{\alpha}}^{(o)} \tag{2.2.2}$$

Since neither the mass of a molecule nor its translational speed are dependent upon the molecule's internal quantum state it follows that the one-particle distribution function for a chemical species α (say) can be derived by simply summing (2.2.1 or 2) over all $\underline{\alpha}$-values that pertain to the chemical species α. In other words

$$f_\alpha = n_\alpha \bar{f}_\alpha^{(o)} \tag{2.2.3}$$

where

$$\bar{f}^{(o)}_{\alpha} = \left(\frac{m_\alpha}{2\pi kT_1}\right)^{3/2} \exp\{-m_\alpha V_{\alpha k} V_{\alpha k}/2kT_1\}, \tag{2.2.4}$$

$V_{\alpha j}$ and m_α are synonymous with $V_{\underline{\alpha}j}$ and $m_{\underline{\alpha}}$, and

$$n_\alpha = \sum_{\underline{\alpha}_\alpha} n_{\underline{\alpha}} \, . \tag{2.2.5}$$

$\sum_{\underline{\alpha}_\alpha}$ implies summation over all quantum states that pertain to the chemical species α.

Clearly one could sum $f^{(o)}_{\underline{\alpha}}$ over any desired set of quantum numbers (say all those for molecular rotation) for a given chemical species so (2.2.5) may refer to all of the chemical species α molecules, regardless of rotational state, that are in a particular vibrational energy level, for example.

2.3 The 'Reaction' Terms

If only binary inelastic collisions are important the results in (2.1.1 and 2) can be used to estimate $\dot{N}_{\underline{\alpha}}$. In view of the remarks in §2.2 about $f_{\underline{\alpha}}$ a first estimate will follow from the use of (2.2.1 and 2); writing this estimate as $\dot{N}^{(o)}_{\underline{\alpha}}$ it can be expressed in the form

$$\dot{N}^{(o)}_{\underline{\alpha}} = \sum_{\underline{\beta}\underline{\gamma}\underline{\delta}}\sum\sum \{n_{\underline{\gamma}} n_{\underline{\delta}} P^{\underline{\alpha}\underline{\beta}}_{\underline{\gamma}\underline{\delta}} - n_{\underline{\alpha}} n_{\underline{\beta}} P^{\underline{\gamma}\underline{\delta}}_{\underline{\alpha}\underline{\beta}}\}, \tag{2.3.1}$$

where

$$P^{\underline{\gamma}\underline{\delta}}_{\underline{\alpha}\underline{\beta}} \equiv \int\!\!\int\!\!\int\!\!\int f^{(o)}_{\underline{\alpha}} f^{(o)}_{\underline{\beta}} g_{\underline{\alpha}\underline{\beta}} I^{\underline{\gamma}\underline{\delta}}_{\underline{\alpha}\underline{\beta}} \sin\theta \, d\theta \, d\phi \, dv_{\underline{\beta}} \, dv_{\underline{\alpha}}, \tag{2.3.2}$$

and similarly for the other reverse 'reaction' probability. Clearly $P^{\underline{\gamma}\underline{\delta}}_{\underline{\alpha}\underline{\beta}}$ is an average value of the total collision cross-section taken over all values of the pre-collision velocities; for obvious reasons they are often called thermally-averaged probabilities.

Under equilibrium conditions the $\{\ \}$ terms in (2.3.1) must all vanish and this enables one to associate the 'forward' and 'reverse' probabilities under these conditions; the ensuing relationship is often also used under non-equilibrium conditions, provided that these are not too-far removed from equilibrium. The existence of a dynamic balance between forward and reverse processes in an equilibrium state provides direct kinetic-theory evidence for the classical chemical "Law of Mass Action".

Actually (2.3.1) is rather a general result in the context of the 'quantised' internal-state model described in §1.1. It forms the basis of most studies of, and expressions for, the 'reaction' terms $\dot{N}_{\underline{\alpha}}$ whether these originate in the relaxation of internal molecular-energy states or in the more gross changes that take place during the progress of chemical reactions.

2.4 Rotational and Vibrational Relaxation; Chemical Reactions

Polyatomic molecules can store energy in rotations about axes through the molecule's mass-centre as well as in vibrations of the two or more atoms that go to make up the particular chemical character of each particular molecule. As a consequence it is necessary to model changes in both internal-molecular energy content and in chemical character. If all of the cross-sections are known (2.3.1) will provide the requisite value of $\dot{N}_{\underline{\alpha}}^{(o)}$, but it is clear that information at the level implicit in the $\underline{\alpha}$-notation, even if all of the cross-sections are known, is usually far too detailed for most purposes.

Consider the following particular case, based on what is usually called nearest-neighbour transition. The probability of a quantum jump from anything other than one level to or from its immediate neighbour is assumed to be zero so that, for a given $\underline{\alpha}$,

$$P_{\underline{\alpha}\underline{\beta}}^{\underline{\gamma}\underline{\beta}} \text{ and } P_{\underline{\gamma}\underline{\beta}}^{\underline{\alpha}\underline{\beta}} = 0, \; \underline{\gamma} \neq \underline{\alpha} - 1, \; \underline{\alpha}, \; \underline{\alpha} + 1. \qquad (2.4.1)$$

It has been assumed in (2.4.1) that the collision partner originally in the state $\underline{\beta}$ remains in that state; going further we assume that collisions in which both partners suffer a change in internal quantum state (so that $\underline{\delta} \neq \underline{\beta}$) have zero probability of occurrence. Then (2.3.1) can be written

$$\dot{\underline{N}}_{\underline{\alpha}}^{(o)} = \sum_{\underline{\beta}} n_{\underline{\beta}} \{ n_{\underline{\alpha}-1} P_{\underline{\alpha}-1}^{\underline{\alpha}} + n_{\underline{\alpha}} P_{\underline{\alpha}}^{\underline{\alpha}} + n_{\underline{\alpha}+1} P_{\underline{\alpha}+1}^{\underline{\alpha}} - n_{\underline{\alpha}} P_{\underline{\alpha}}^{\underline{\alpha}-1} - n_{\underline{\alpha}} P_{\underline{\alpha}}^{\underline{\alpha}} - n_{\underline{\alpha}} P_{\underline{\alpha}}^{\underline{\alpha}+1} \} , \quad (2.4.2)$$

where the sub- and superscripts $\underline{\beta}$ have been omitted for brevity, and the influence of the elastic-collisions, that depend upon $P_{\underline{\alpha}}^{\underline{\alpha}}$, cancel out as they must.

Nearest-neighbour transitions occur between the non-degenerate energy-levels $\epsilon_{\underline{\alpha}}$ (per molecule) in a simple-harmonic vibrator for which, ignoring the zero-point energy,

$$\epsilon_{\underline{\alpha}} = \underline{\alpha} h\omega$$

where h (in this Section only) is Planck's constant, ω is the vibrator's frequency and $\underline{\alpha}$ here is the number, 0,1,2, etc, that signifies the number of energy level in the vibrator. The thermally averaged transition probabilities for excitation and de-excitation by translation-vibration exchange in the simple-harmonic vibrator system have two important additional features, namely

$$P_{\underline{\alpha}}^{\underline{\alpha}} = \underline{\alpha} P_1^{o} \quad ; \quad P_{\underline{\alpha}}^{\underline{\alpha}+1} = P_{\underline{\alpha}}^{\underline{\alpha}} \quad \exp(h\omega/kT_1) \quad (2.4.4a,b)$$

where the value 0 for $\underline{\alpha}$ denotes the lowest, or ground, state of the oscillator. Installing these relationships into (2.4.2) gives

$$\dot{N}_{\underline{\alpha}}^{(o)} = nP_1^{o} \{ \underline{\alpha} n_{\underline{\alpha}-1} e^{-h\omega/kT_1} - \underline{\alpha} n_{\underline{\alpha}} + (\underline{\alpha} + 1) n_{\underline{\alpha}+1} - (\underline{\alpha} + 1) n_{\underline{\alpha}} e^{-h\omega/kT_1} \} \quad (2.4.5)$$

where $\Sigma_{\underline{\beta}} n_{\underline{\beta}}$ has been set equal to n, the total number of molecules per unit volume. Evidently one must solve all of the 'species'-$\underline{\alpha}$ equations simultaneously, even in the present ultra-simple situation, if one is to keep track of all of the level populations in the vibrator. However, if each $\dot{N}^{(o)}_{\underline{\alpha}}$ is multiplied by the energy $\underline{\alpha}\hbar\omega$ in that particular quantum level, and the whole then summed over all $\underline{\alpha}$, the result will be the rate of gain of energy in the whole vibrational mode per unit volume. After a little re-arrangement it can be shown that

$$\sum_{\underline{\alpha}=0} \varepsilon_{\underline{\alpha}} \dot{N}^{(o)}_{\underline{\alpha}} = nP^o_1 (1 - e^{-\hbar\omega/kT_1})\left\{\frac{n\hbar\omega}{(\exp(\hbar\omega/kT_1) - 1)} - \sum_{\underline{\alpha}=0} \varepsilon_{\underline{\alpha}} n_{\underline{\alpha}}\right\} . \quad (2.4.6)$$

There are several things to remark upon in (2.4.6). First, nP^o_1 has the dimensions of $(\text{time})^{-1}$ and it is convenient to define

$$\tau^{vib} \equiv \left[nP^o_1 (1 - e^{-\hbar\omega/kT_1})\right]^{-1} = \tau^{vib}(p,T_1) , \quad (2.4.7)$$

where τ^{vib} is a relaxation time for the vibrational mode. Second

$$\sum_{\underline{\alpha}=0} \varepsilon_{\underline{\alpha}} n_{\underline{\alpha}} = \rho e^{vib} \quad (2.4.8)$$

is the local vibrational energy in unit volume, and third,

$$n\hbar\omega(\exp(\hbar\omega/kT_1) - 1)^{-1} \equiv \rho e^{vib}_e(p,T_1) \quad (2.4.9)$$

is the energy in unit volume in a vibrational mode that is in equilibrium at the local translational temperature T_1. It follows that

$$\sum_{\underline{\alpha}=0} \varepsilon_{\underline{\alpha}} \dot{N}^{(o)}_{\underline{\alpha}} = \frac{\rho}{\tau^{vib}}\left[e^{vib}_e - e^{vib}\right] , \quad (2.4.10)$$

which shows that local changes in vibrational energy per unit mass proceed at rates that are directly proportional to the difference between local actual and local equilibrium energy values. The rate of the process is governed by τ^{vib} which can be seen from (2.4.7) to be inversely proportional to nP_1^o. From its definition in (2.3.2) it is clear that nP_1^o must be some fraction of the molecular collision frequency; calculations show that it is a very small fraction (e.g. Stupochenko, et al, (1967), Chaps, 2 & 4; Clarke & McChesney, (1976), Chaps, 3 & 4) so that if τ^{coll} is the reciprocal of the collision frequency it is invariably true that

$$\tau^{coll} \ll \tau^{vib} . \tag{2.4.11}$$

The developments leading to (2.4.10) are very basic in concept and deal only with the relaxation of molecules treated as simple-harmonic vibrators, with energy transfer taking place via exchanges of vibrational and translational energies. Vibrational relaxation is complicated, even in a pure chemical-component gas, by the population and de-population of a molecule's vibrational energy levels through interaction of molecular vibration with both the translational motion and with the vibrational motion of a molecule's collision partner. The vibration-vibration interactions can have particular influence over the rates of population of lower vibrational levels in real anharmonically vibrating molecules, and so can result in some complicated vibrational-mode behaviour in response to large changes in local translational energy levels. The review by Rich and Treanor (1970) is particularly useful.

Rotational relaxation is much more difficult to deal with than even the adjustment of vibrational modes. Discussions can be found in the books just quoted and amount to the fact that (2.4.11) can be augmented to read

$$\tau^{coll} < \tau^{rot} << \tau^{vib} , \tag{2.4.12}$$

with τ^{rot} usually about 5 to 10 times τ^{coll}.

If the description of the gas implied in the subscripts $\underline{\alpha}$, etc, is coarsened as described in §2.2 to the extent that α (without an underbar) denotes a chemical species, regardless of its quantum state, (2.3.1) can be interpreted to mean the rate of production of numbers of molecules per unit volume per unit time in a chemical reaction. For example, in a reaction

$$A_\gamma + A_\delta \rightleftarrows A_\alpha + A_\beta \tag{2.4.13}$$

the rate of appearance of chemical species α in molecules per unit volume per unit time is given by

$$n_\gamma n_\delta P^{\alpha\beta}_{\gamma\delta} - n_\alpha n_\beta P^{\gamma\delta}_{\alpha\beta} . \tag{2.4.14}$$

This sort of expression is usually written with $\tilde{k}_f$ in place of $P^{\alpha\beta}_{\gamma\delta}$ and $\tilde{k}_r$ in place of $P^{\gamma\delta}_{\alpha\beta}$, where $\tilde{k}_{f,r}$ are called specific reaction rate 'constants'. From what we have done so far it is evident that $\tilde{k}_f$, $\tilde{k}_r$ will in general depend upon T_1 since all internal modes are tacitly presumed to have energy-contents given by the local pseudo-equilibrium implicit in the use of T_1 .

The link between $\tilde{k}_{f,r}$ quantities and molecular collisions is clear. We shall pick up such matters again later on in special circumstances and meanwhile we conclude this Section with the comment that internal-mode relaxations and chemical reactions are so complicated in general that they are treated as physical processes, fit for study in their own right, (see e.g. Clarke & McChesney, 1976, Chaps. 3, 4 and 5) with usually little or no reference to Boltzmann's equation.

2.5 Fluxes and Affinities

This is neither the time nor the place to recapitulate the details or the implications of the Chapman-Enskog solution of Boltzmann's equation. Many references deal at length with the case of monatomic gases (e.g. Chapman & Cowling, 1970; Hirschfelder, et al, 1954; Vincenti & Kruger, 1965, provide most useful guidance through the complexities inherent in Chapman-Enskog solutions) and also describe some of the modifications that are necessitated by the extension to polyatomic gases. A complete formal kinetic theory of polyatomic gases was first given by Wang Chang, Uhlenbeck & de Boer (1964) and is still the most widely quoted paper on this topic. For present purposes, all that we need is a link between the fluxes $g_{\underline{\alpha}j}$, p_{ij} and q_j and the effects, or affinities, that drive them. A short summary of the processes involved in the specific case of polyatomic molecules has been given by Clarke & McChesney (1976; §1.7); salient features of this summary are given here.

First it must be recognized that any affinity, in the shape of spatial (vector) gradients of $n_{\underline{\alpha}}$, p & T_1, of the rate-of-strain tensor $\partial u_i/\partial x_j$ <u>and</u> of the scalar 'reaction' terms $\dot{N}_{\underline{\alpha}}$ is <u>a priori</u> capable of driving any of the fluxes. Limiting one's attention to isotropic media simplifies the cross-couplings between fluxes and affinities, so that $q_{\underline{\alpha}j}$ is driven by spatial gradients of $n_{\underline{\alpha}}/n$, p & T_1, while q_j is similarly dependent (cf (1.2.13 and 14) and the role of diffusion velocity $u_{\underline{\alpha}j}$ as transporter of $e_{\underline{\alpha}}^{int}$).

In principle the stress tensor τ_{ij} (see (1.3.9)) depends upon both $\partial u_i/\partial x_j$ and $\dot{N}_{\underline{\alpha}}$. With the sort of easy interchange of energy that occurs between translation and rotation (cf §2.4) the related $\dot{N}_{\underline{\alpha}}$ terms are often said to "give rise" to a bulk or volume viscosity. Since it can be shown, for any mixture that can be treated as described in §2.2 and 3, that the actual contribution of $\dot{N}_{\underline{\alpha}}$ to τ_{ij} is zero such a statement can be misleading. The matter will be referred to again in §3.6 below.

The fluxes exist primarily as a result of one basic affinity, $g_{\underline{\alpha}j}$ primarily from $\partial(n_{\underline{\alpha}}/n)/\partial x_j$ and τ_{ij} from $\partial u_j/\partial x_i$; q_j in a polyatomic gas must take account of both temperature gradient $\partial T_1/\partial x_j$ and diffusion-velocity drives. Such cross-transport influences as the Soret effect (thermal diffusion) on diffusion velocities and the Dufour effect (reciprocal to thermal diffusion) on energy-flux can be crucial in special circumstances, but are usually negligible in most gas-dynamical or flow processes. All of these effects are fully described in the various books that have been mentioned so far; one should mention, in addition, the very directly practical text by Bird, Stewart & Lightfoot (1960) as well as the useful summaries and sets of ideas described in the Appendices to the book by Williams (1985). Listing only the direct flux/affinity links, we shall find the following relationships useful in our discussions of gas-dynamical matters;

$$\rho g_{\underline{\alpha}j} = m_{\underline{\alpha}} n^2 \sum_{\underline{\beta}} m_{\underline{\beta}} D_{\underline{\alpha}\underline{\beta}} \frac{\partial}{\partial x_j}(n_{\underline{\beta}}/n) \quad , \; D_{\underline{\alpha}\underline{\alpha}} = 0 \quad ; \tag{2.5.1}$$

$$\tau_{ij} = \eta\{\frac{\partial u_j}{\partial x_i} + \frac{\partial u_i}{\partial x_j}\} - \frac{2}{3}\eta\frac{\partial u_k}{\partial x_k}\delta_{ij} \quad ; \tag{2.5.2}$$

$$q_j = -\lambda\frac{\partial T_1}{\partial x} + \sum_{\underline{\alpha}} g_{\underline{\alpha}j} h_{\underline{\alpha}} \; , \qquad h_{\underline{\alpha}} = \frac{5}{2}\frac{p_{\underline{\alpha}}}{\rho_{\underline{\alpha}}} + e^{int}_{\underline{\alpha}} \; , \tag{2.5.3}$$

where $h_{\underline{\alpha}}$ is the enthalpy per unit mass of 'species' $\underline{\alpha}$. The new symbols that appear in these equations denote the so-called transport coefficients, namely $D_{\underline{\alpha}\underline{\beta}}$, a multicomponent diffusion coefficient, η the shear viscosity and λ the thermal conductivity for the gas mixture. All of $D_{\underline{\alpha}\underline{\beta}}$, η and λ are ultimately dependent on the collision integrals in the Boltzmann equation (1.1.1) (see also §2.1), and hence upon the collision cross-sections. In view of the remarks made in §2.2 about the infrequency of inelastic collisions it is not unusual to assume that the transport coefficients can be adequately predicted by an elastic-collision theory that makes use of the idea of symmetrical inter-molecular short-range force fields. The results of doing so are very satisfactory, especially when some overtly polyatomic behaviour, to be described below, is taken into account. A fairly full account of the influence of inelastic collisions on Chapman-Enskog solutions of the Boltzmann equation for molecules of the same chemical species with internal energy modes is given by Kogan (1969, §3.10).

3. DYNAMICS OF 'REACTIVE' GASES

3.1 Equations

It has already been shown how appropriate moments of the Boltzmann equation lead to equations of conservation of mass (1.3.4), momentum (1.3.8) and energy (1.3.10). As written down in §1.3 these equations provide no direct evidence of the role of 'reactions'. However, intrinsic energy e depends directly on $n_{\underline{\alpha}}$, as is evident from (1.2.10), and so one cannot avoid the need to solve the 'species' equations simultaneously with the usual conservation equations. The intrinsic complexity of this exercise has been revealed in §2.4 (see (2.4.5) for example), but that sub-section also suggests a possible way out of this particular dilemma. Equation (2.4.10) shows that the 'reaction' source term reduces to a simple form when it is overall, as opposed to highly detailed, information that is sought. In that case the overall as opposed to detailed quantity that is sought is the energy per unit mass in the whole vibrational mode. However, it must be said that (2.4.10) is actually a rather unusual result insofar as it relates the relaxation time τ^{vib} to molecular transition probabilities in the elegantly simple fashion revealed by (2.4.7). This matter will be referred to again below but, meanwhile, we should note the effect of multiplying (1.3.1) by the energy-content in level-$\underline{\alpha}$ per molecule, namely

$$\varepsilon_{\underline{\alpha}} = m e_{\underline{\alpha}}^{int} , \qquad m_{\underline{\alpha}} = m \text{ for all } \underline{\alpha} \tag{3.1.1}$$

(cf §1.1 and 2.4), followed by summation over all $\underline{\alpha}$. The result is

$$\frac{\partial}{\partial t}(\rho e^{int}) + \frac{\partial}{\partial x_k}(\rho u_k e^{int}) + \frac{\partial}{\partial x_k}(\Sigma_{\underline{\alpha}} g_{\underline{\alpha}} e_{\underline{\alpha}}^{int}) = \Sigma_{\underline{\alpha}} \varepsilon_{\underline{\alpha}} \dot{N}_{\underline{\alpha}} , \tag{3.1.2}$$

where

$$\sum_{\underline{\alpha}} \rho_{\underline{\alpha}} e_{\underline{\alpha}}^{int} = \rho e^{int} \qquad (3.1.3)$$

in analogy with (2.4.8).

Also by analogy with (2.4.10), the right-hand side of (3.1.2) will be written as

$$\frac{\rho}{\tau} [e_e^{int} - e^{int}] \quad ,$$

and it is clear that one must now deal with the third term in (3.1.2) that involves diffusive mass fluxes $g_{\underline{\alpha}j}$.

Note that the quantity

$$\sum_{\underline{\alpha}} e_{\underline{\alpha}}^{int} g_{\underline{\alpha}j} = \sum_{\underline{\alpha}} \rho_{\underline{\alpha}} u_{\underline{\alpha}j} e_{\underline{\alpha}}^{int} = \sum_{\underline{\alpha}} m n_{\underline{\alpha}} e_{\underline{\alpha}}^{int} u_{\underline{\alpha}j} \qquad (3.1.4)$$

also appears in the expression (2.5.3) for q_j. When m is the same for all molecules in the mixture (2.5.1) reduces to

$$n_{\underline{\alpha}} u_{\underline{\alpha}j} = n \sum_{\underline{\beta} \neq \underline{\alpha}} D_{\underline{\alpha}\underline{\beta}} \frac{\partial}{\partial x_j} (n_{\underline{\beta}}/n) \quad , \qquad (3.1.5)$$

and if the reasonable assumption is made that all diffusion coefficients are the same, so that

$$D_{\underline{\alpha}\underline{\beta}} = \tilde{D} \text{ for all } \underline{\alpha}, \underline{\alpha} \neq \underline{\beta} \quad , \qquad (3.1.6)$$

it follows that

$$n_{\underline{\alpha}} u_{\underline{\alpha}j} = -n \tilde{D} \frac{\partial}{\partial x_j} (n_{\underline{\alpha}}/n) \quad . \qquad (3.1.7)$$

Thus

$$\sum_{\underline{\alpha}} e_{\underline{\alpha}}^{int} g_{\underline{\alpha}j} = -\rho \tilde{D} \frac{\partial}{\partial x_j} e^{int}. \qquad (3.1.8)$$

In view of the fact that $p_{\underline{\alpha}}/\rho_{\underline{\alpha}}$ is equal to 3/2 times $kT_{\underline{\alpha}1}/m_{\underline{\alpha}}$ (cf (1.2.19)), that $T_{\underline{\alpha}1}$ is the same as T_1 for all $\underline{\alpha}$, and that $m_{\underline{\alpha}}$ is similarly equal to m, it is clear that $p_{\underline{\alpha}}/\rho_{\underline{\alpha}}$ is independent of $\underline{\alpha}$ in the present model. As a result the second term in (2.5.3) reduces to $\sum_{\underline{\alpha}} g_{\underline{\alpha}j} e_{\underline{\alpha}}^{int}$ and so (3.1.8) shows that

$$q_j = -\lambda \frac{\partial T_1}{\partial x_j} - \rho \tilde{D} \frac{\partial}{\partial x_j} e^{int} . \tag{3.1.9}$$

For the same model (3.1.2) becomes

$$\frac{\partial}{\partial t} (\rho e^{int}) + \frac{\partial}{\partial x_k}(\rho u_k e^{int}) - \frac{\partial}{\partial x_k} (\rho \tilde{D} \frac{\partial e^{int}}{\partial x_k}) = \frac{\rho}{\tau}[e_e^{int} - e^{int}] \tag{3.1.10}$$

and it is important to note that (1.2.10) gives

$$e = e^{tr} + e^{int} = \frac{3}{2} \frac{k}{m} T_1 + e^{int} \equiv C_{vf} T_1 + e^{int} \tag{3.1.11}$$

for the intrinsic energy. The last equation in (3.1.11) defines the translational specific heat at constant volume; subscript f is used to denote a 'frozen' state for reasons that will become apparent.

3.2 A General Model

The five conservation equations (1.3.4, 8 and 10) have as dependent variables ρ, u_j, p, τ_{ij} , e and q_j , which total eighteen (separate quantities) in all. Equations (2.5.2), (3.1.9, 10 and 11) add another fourteen equations in all, but at the expense of introducing two more quantities namely T_1 and e^{int}. The system is closed by means of (1.2.18) which relates p, ρ(=mn) and T_1. Thus, although the system of equations is now complete (given that all of the 'molecular-collision' effects η, $\tilde{D}$, λ and τ are known of course!), it is certainly quite complicated even though it has been derived (particularly (3.1.9), 10 and 11) for a special model of a simple relaxing gas.

That the system as derived so far can be extended to apply to a rather broader class of phenomena than pure relaxation should be clear from what has been done so far. Details can be found in the work of Vincenti & Kruger (1965), Clarke & McChesney (1964, 1976) and Wegener

(1969). With the aid of a new variable q, called the non-equilibrium variable, the equations can be generalised somewhat, as follows

$$\frac{\partial\rho}{\partial t} + \frac{\partial}{\partial x_k}(\rho u_k) = 0 \quad , \tag{3.2.1}$$

$$\frac{\partial}{\partial t}(\rho u_j) + \frac{\partial}{\partial x_k}\{p\delta_{jk} - \tau_{jk} + \rho u_k u_j\} = 0 \ , \ j = 1,2,3 \ , \tag{3.2.2}$$

$$\frac{\partial}{\partial t}[\rho(h + \tfrac{1}{2}u^2)] + \frac{\partial}{\partial x_k}[\rho u_k(h + \tfrac{1}{2}u^2) + q_k - u_j\tau_{jk}] - \frac{\partial p}{\partial t} = 0 \ , \tag{3.2.3}$$

$$\frac{\partial}{\partial t}(\rho q) + \frac{\partial}{\partial x_k}[\rho u_k q - \rho\tilde{D}\frac{\partial q}{\partial x_k}] = w(p,\rho,q) \quad , \tag{3.2.4}$$

$$\tau_{ij} = \eta[\frac{\partial u_j}{\partial x_i} + \frac{\partial u_i}{\partial x_j}] - \frac{2}{3}\eta\frac{\partial u_k}{\partial x_k}\delta_{ij} \quad , \quad i,j = 1,2,3, \tag{3.2.5}$$

$$q_j = -\lambda\frac{\partial T_1}{\partial x_j} - \rho\tilde{D}\frac{\partial q}{\partial x_j} \quad , \quad j = 1,2,3 \quad , \tag{3.2.6}$$

$$p = \rho\tilde{R}(q)T_1 \quad , \tag{3.2.7}$$

$$e = e_1(T_1, q) + q \quad , \tag{3.2.8}$$

$$h = e + p/\rho = h(p,\rho,q) \ . \tag{3.2.9}$$

There is a total of twenty-four variables not counting $\tilde{D}$, η and λ in these twenty-one equations; however, it is assumed that $\tilde{D}$, η and λ are known functions of local conditions, together with e_1 (the 'active' intrinsic energy), $\tilde{R}$ (the gas constant) and w (the non-equilibrium source term; by analogy with (3.1.10) the reaction, or relaxation, time resides in the reaction source term w).

3.3 Acoustic Waves; Linear Equations

An examination of the way in which weak plane (acoustic) waves propagate through a reactive gas enables one to illustrate several features of the behaviour of these substances. To do this it is necessary to linearise the equations described in §3.2. The energy and 'relaxation' equation (3.2.3 and 4), respectively, require some special treatment. Starting with the latter, note that when q has an equilibrium value q_e at a given p and ρ the source or reaction term w is zero:

$$w(p,\rho,q_e) = 0 \Rightarrow q_e = q_e(p,\rho) . \qquad (3.3.1)$$

The last result follows from the implicit-function theorem provided that $\partial w/\partial q$ does not vanish for the given p and ρ.

Assuming that the unperturbed system (denoted by $(\ )_o$) is in a state of equilibrium, Taylor's theorem can be used to show that

$$w(p,\rho,q) \approx \frac{\rho_o}{\tau_o}(q_e - q) , \qquad (3.3.2)$$

to leading order, where the relaxation time τ_o is given in general by

$$\tau_o = - \rho_o/[\partial w(p_o, \rho_o, q_o)/\partial q] . \qquad (3.3.3)$$

Comparison of this purely mathematical result with the physical derivation of (2.4.10), for example, is interesting and recall the remarks made prior to (3.1.1).

The linear version of (3.2.4) is therefore

$$\tau_o q_t - \tau_o \rho_o \tilde{D}_o q_{xx} + q - q_{epo} p - q_{e\rho o} \rho = 0 , \qquad (3.3.4)$$

where q_{ep} is $\partial q_e/\partial p$, $q_{e\rho}$ is $\partial q_e/\partial \rho$ and the $(\ )_o$ indicates evaluation in the equilibrium unperturbed atmosphere; p,ρ,q all signify <u>perturbations</u> from ambient values p_o, ρ_o, q_o in (3.3.4).

The linear energy equation (3.2.3) is

$$\rho_o h_t - \lambda_o T_{1xx} - \rho_o \tilde{D}_o q_{xx} - p_t = 0 \ , \qquad (3.3.5)$$

but it is necessary to eliminate h and T_1 in favour of p,ρ and q in order to make progress with the analysis. It is easier to do this in general terms than it is to choose special forms for h and T_1 as functions of p,ρ and q.

For brevity later on it will be assumed that T_1 does not depend upon q; this limits our model to that of a pure relaxing gas or, at least, to one for which no changes of molecular weight occur.

Rearrangement of (3.3.5) is perfectly straightforward, but does rely on the use of a number of subsidiary relationships amongst thermodynamical quantities which are listed here for convenience:

$$\begin{aligned}
&\rho\alpha_f = -(\partial\rho/\partial T_1)_{p,q} \ , \quad \rho\beta_f = (\partial\rho/\partial p)_{T_1,\,q} \ ; \\
&C_{pf} = (\partial h/\partial T_1)_{p,q} \ , \quad C_{vf} = (\partial e/\partial T_1)_{\rho,q} \ , \qquad (3.3.6) \\
&\gamma_f = C_{pf}/C_{vf} \ ; \quad a_f^2 = -(\partial h/\partial\rho)_{p,q}/(\partial e/\partial p)_{\rho,q} = \gamma_f/\rho\beta_f \ ; \\
&(\partial h/\partial\rho)_{p,q} = -C_{pf}/\rho\alpha_f \ .
\end{aligned}$$

With the help of (3.3.6), (3.3.5) appears in the form

$$-a_{fo}^{-2}\{p_t - \gamma_{fo}\varkappa_{fo}p_{xx}\} + \{\rho_t - \varkappa_{fo}\rho_{xx}\} + (h_{\rho o})^{-1}\{q_t - \tilde{D}_o q_{xx}\} = 0 \ , \qquad (3.3.7)$$

where

$$\varkappa_f \equiv \lambda/\rho C_{pf} \qquad (3.3.8)$$

is the thermal diffusivity based on properties determined with q fixed or frozen. The shorthand notation h_ρ has been used for $(\partial h/\partial\rho)_{p,q}$

Together with

$$\rho_t + \rho_o u_{xx} = 0 \quad , \quad \rho_o u_t + p_x - \frac{4}{3}\eta_o u_{xx} = 0 \quad , \qquad (3.3.9a,b)$$

which are the linear versions of (3.2.1) and (3.2.3) in one space dimension, (3.3.5 and 7) now make up four equations for the four variables p, ρ, u and q.

3.4 Acoustic Waves: Dispersion Relation

Assume that the acoustic waves are harmonic in the sense that

$$\psi = \tilde{\psi} \exp[i\Phi(x,t)] \quad ; \quad \psi = p, \rho, u, q, \qquad (3.4.1)$$

where $\tilde{\psi}$ is a constant amplitude number for each variable, and the phase Φ defines a frequency ω and wave-number k, complex in general, via

$$\omega = \Phi_t \quad , \quad -k = \Phi_x \quad . \qquad (3.4.2)$$

Substituting (3.4.1) into (3.3.4, 7, 9a and b) leads to a set of four simultaneous homogeneous linear algebraic equations for the amplitudes $\tilde{\psi}$. The need for the determinant of the coefficients of $\tilde{\psi}$ to vanish for consistency of these algebraic equations gives a relation (the <u>dispersion relation</u>) between ω and k. The relationship that emerges is lengthy and involves a number of different physical quantities. In view of (3.3.4) q_{ep} and $q_{e\rho}$ both appear, so that it is important to augment the 'frozen' thermodynamic relations (3.5.6) with some 'equilibrium' counterparts. Writing

$$h_e = h(p,\rho,q_e(p,\rho)) \quad , \quad e_e = e(p,\rho,q_e(p,\rho)) \quad , \qquad (3.4.3)$$

the equilibrium analogue of a_f^2 is a_e^2 ,

$$a_e^2 = -\left(\frac{\partial h_e}{\partial \rho}\right)_p / \left(\frac{\partial e_e}{\partial p}\right)_\rho = -[h_\rho + q_{e\rho}]/[e_p + q_{ep}] \quad , \qquad (3.4.4)$$

since $(\partial h/\partial q)_{p,\rho}$ is unity in the present case (see (3.2.8 and 9).

Thus

$$a_e^2/a_f^2 \equiv 1/a^2 = [h_\rho + q_{e\rho}]/[h_\rho - a_f^2 q_{e\rho}] < 1 \ . \tag{3.4.5}$$

The inequality follows from arguments based on thermodynamic stability (e.g. Clarke & McChesney, 1976, §1.17). With

$$\left(\frac{\partial h_e}{\partial \rho}\right)_p = -\frac{C_{pe}}{\rho\alpha_e} \ , \tag{3.4.6}$$

where

$$C_{pe} = \left(\frac{\partial h_e}{\partial T_1}\right)_p \ , \qquad \rho\alpha_e = -\left(\frac{\partial \rho_e}{\partial T_1}\right)_p \ ,$$

it can be shown that

$$q_{ep}/h_\rho = a_f^{-2}[1 - a^2(C_{pe}\,\alpha_f/C_{pf}\,\alpha_e)] \equiv a_f^{-2}[1 - a^2 r^{-1}] \ ,$$

$$q_{e\rho}/h_\rho = [(C_{pe}\,\alpha_f/C_{pf}\,\alpha_e) - 1] \equiv [r^{-1} - 1] \ . \tag{3.4.7}$$

If one further defines

$$\tau^* = r\tau_o \tag{3.4.8}$$

it eventually turns out that the dispersion relation is

$$\omega^2\left(1 - i\frac{4}{3}\nu_o\frac{k^2}{\omega}\right)\left\{(i\omega\tau^* + a^2) - i\frac{k^2}{\omega}\left[(i\omega\tau^* + r)\gamma_{fo}\varkappa_{fo} + (i\omega\tau^* + a^2 - r)\tilde{D}_o\right]\right\}$$
$$= a_{fo}^2 k^2\left\{(i\omega\tau^* + 1) - i\frac{k^2}{\omega}\left[(i\omega\tau^* + r)\varkappa_{fo} + (i\omega\tau^* + 1 - r)\tilde{D}_o\right]\right\} \ , \tag{3.4.9}$$

which might be reasonably compact but is hardly very informative as it stands!

First note that the terms $k^2[\,]/\omega$ in (3.4.9) are all like $k^2\varkappa_{fo}/\omega$ in magnitude, since $\tilde{D}_o$ and $\varkappa_{fo}$ do not differ radically; the viscous term $4\nu_o k^2/3\omega$ is similar.

It will be assumed that ω is real while k is complex, so that

$$k = k_r + ik_i \ ; \quad i\Phi = i(\omega t - k_r x) + k_i x \ . \tag{3.4.10}$$

The modulus of k will be well approximated by k_r and the phase velocity

a_{po} is evidently

$$a_{po} = \omega/k_r \ . \qquad (3.4.11)$$

Then $k^2 \varkappa_{fo}/\omega$ is of the same order as $\omega\varkappa_{fo}/a_{po}^2$; but any diffusion coefficient is of the order of a mean molecular thermal speed squared times τ^{coll} , the mean molecular collision time. Since a_{po} is itself of the order of a mean molecular thermal speed it can be seen that $\omega\varkappa_{fo}/a_{po}^2$ is of order $\omega\tau^{coll}$ which should be small if one is not to violate the principles implicit in the Chapman-Enskog solution of Boltzman's equation (cf §2.2). As a consequence (3.4.9) can be written sufficiently accurately as

$$a_{fo}\, k \simeq \omega\left(\frac{a^2 + i\omega\tau^*}{1 + i\omega\tau^*}\right)^{\frac{1}{2}} \left\{1 - i\,\frac{k^2}{2\omega}\left[\frac{4}{3}\nu_o + \varkappa_o^i\right] . . .\right\} \ , \qquad (3.4.12)$$

where

$$\varkappa_o^i \equiv \frac{(i\omega\tau^* + r)[i\omega\tau^*(\gamma_{fo} - 1) + (\gamma_{fo} - a^2)]\varkappa_{fo} + (a^2 - 1)r\tilde{D}_o}{(i\omega\tau^* + 1)(i\omega\tau^* + a^2)} \ . \qquad (3.4.13)$$

3.5 Acoustic Waves: Relaxation Effects

It is clear from (3.4.12) that relaxation effects are fundamentally different from those of diffusion (in the general sense implicit in $\frac{4}{3}\nu_o + \varkappa_o^i$). Therefore consider the relaxation effects in isolation, to which end (3.4.12) is simplified to read

$$a_{fo}\, k \simeq \omega m e^{-i\Theta} \ , \qquad (3.5.1)$$

where

$$m = \left(\frac{a^4 + (\omega\tau^*)^2}{1 + (\omega\tau^*)^2}\right)^{\frac{1}{4}} \ , \qquad (3.5.2)$$

$$\Theta = \tfrac{1}{2}\tan^{-1}\left(\frac{(a^2 - 1)\omega\tau^*}{a^2 + (\omega\tau^*)^2}\right) \simeq \tfrac{1}{2}(a^2 - 1)\left(\frac{\omega\tau^*}{a^2 + (\omega\tau^*)^2}\right) \qquad (3.5.3)$$

The last approximation here is permissible because the ratio a^2 exceeds unity by only small amounts; in fact $\Theta << 1$ is invariably true.

The phase velocity is given by (3.4.11) combined with (3.5.1);

$$a_{po} = a_{fo}/m\cos\Theta \simeq a_{fo}/m \quad . \qquad (3.5.4)$$

Since m varies in the range $a > m > 1$ as $\omega\tau^*$ varies in $0 < \omega\tau^* < \infty$ it follows that

$$a_{eo} < a_{po} < a_{fo} \quad . \qquad (3.5.5)$$

Thus the system is <u>dispersive</u>, with low-frequency ($\omega\tau^* = 0$) waves propagating at the (slower) <u>equilibrium sound speed</u> a_{eo} ; high-frequency ($\omega\tau^* = \infty$) waves propagate at the (faster) <u>frozen sound speed</u>.

That the system is <u>dissipative</u> as well as dispersive, can be seen from (3.4.1 and 10) combined with (3.5.1), since

$$k_i = -\omega m \sin\Theta/a_{fo} \simeq -\omega m\Theta/a_{fo} \quad . \qquad (3.5.6)$$

The wavelength ℓ_w of a sound wave is given by

$$\ell_w = \frac{2\pi}{k_r} = \frac{2\pi a_{fo}}{\omega m \cos\Theta} \simeq \frac{2\pi a_{fo}}{m\omega} \quad , \qquad (3.5.7)$$

so that the 'absorption per wavelength', as it is sometimes called, is

$$\zeta \simeq \ell_w \frac{\omega m\Theta}{a_{fo}} = 2\pi\Theta \simeq \pi(a^2 - 1)\frac{\omega\tau^*}{a^2 + (\omega\tau^*)^2} \quad . \qquad (3.5.8)$$

The maximum value of ζ is $\frac{1}{2}\pi(a - a^{-1})$ when $\omega\tau^*$ is equal to a. This fact can be exploited in the analysis of experimental records, but it is rather more interesting for present purposes to note that (with a_{fo}, a_{eo}, τ^* <u>fixed</u>)

$$k_i \to -\tfrac{1}{2}(1 - a^{-2})(\omega\tau^*/a_{eo}) \to 0 \quad , \quad \omega \to 0 \; , \qquad (3.5.9)$$

$$k_i \to -\tfrac{1}{2}(a^2 - 1)/a_{fo}\tau^* \quad , \quad \omega \to \infty \; . \qquad (3.5.10)$$

As a result, high frequency waves travelling at speeds near a_{fo} in a particular gas are damped by relaxation effects, while low frequency waves travelling at speeds near to a_{eo} are hardly damped at all (but see below).

3.6 Equivalent Bulk Viscosity and Eucken Conductivity

A useful approximate version of the dispersion relation (3.4.12) can be written with the aid (3.5.2, 3 and 4) as follows.

$$k \simeq \frac{\omega}{a_{po}} - i\,\frac{\omega^2}{2a_{po}^3}\left[\frac{2\Theta a_{po}^2}{\omega} + \frac{4}{3}\nu_o + \varkappa_o^i\right], \qquad (3.6.1)$$

where Θ times ν_o and $\varkappa_o^i$ is neglected as an effect of second order.

Evidently viscous effects contribute to the absorption of sound, as do the <u>real</u> parts of $\varkappa_o^i$. Consider (3.4.13) in two limiting cases;

$$\varkappa_o^i \sim (\gamma_{fo} - 1)\varkappa_{fo}, \quad \omega\tau^* \sim \infty, \qquad (3.6.2)$$

$$\varkappa_o^i \sim \frac{r}{a^2}\left[(\gamma_{fo} - a^2)\varkappa_{fo} + (a^2 - 1)\tilde{D}_o\right], \quad \omega\tau^* \sim 0. \qquad (3.6.3)$$

The damping effects of viscosity and heat-conduction at high frequency are therefore summarised in the group $\frac{4}{3}\nu_o + (\gamma_{fo} - 1)\varkappa_{fo}$, which is the classical result for a gas with 'frozen' properties.

Much the most interesting result is in (3.6.3). If, as mentioned in §3.3, T_1 does not depend upon q it can be shown (Clarke & McChesney, 1976, §1.17 and note (3.2.7)) that

$$C_{pe} - C_{ve} = C_{pf} - C_{vf} = R = \text{constant}. \qquad (3.6.4)$$

As a consequence it must be true that $q_e(p,\rho)$ depends upon T_1 alone, so that

$$C_{pe} = C_{pf} + \frac{dq_e}{dT_1} \equiv C_{pf} + C, \qquad (3.6.5)$$

where C is the 'specific heat' of the internal mode. Using the definitions of r and a^2 in §3.4, where we may write

$$a^2 = C_{pf}C_{ve}/C_{pe}C_{vf} \tag{3.6.6}$$

in view of (3.6.4), etc, (3.6.3) can be rearranged to read

$$\varkappa_o^i = (\gamma_{eo} - 1)\varkappa_{eo} = (\frac{C_{peo}}{C_{veo}} - 1)\,\frac{(\lambda_o + \rho_o\tilde{D}_oC_o)}{\rho_oC_{peo}} \tag{3.6.7}$$

Evidently $\lambda + \rho\tilde{D}C$ is an 'equilibrium' thermal conductivity; it shows that under conditions for which changes are slow on the scale of the relaxation times the usual translational thermal conductivity λ must be augmented by a 'diffusive' transport of internal energy that follows the translational temperature gradient. This is Eucken's correction to conductivity to account for rapidly relaxing internal-energy modes (like molecular rotation).

Furthermore, when $\omega\tau^*$ is small the phase velocity is a_{eo} (as described in §3.5) and the 'relaxation damping' factor $2\Theta a^2_{po}/\omega$ in (3.6.1) behaves like

$$2\Theta a^2_{po}/\omega \simeq (1 - a^{-2})a^2_{eo}\tau^* = (p\tau_o\,\frac{RC_o}{C^2_{veo}})\,\frac{1}{\rho_o} \equiv \eta_{vo}\,\frac{1}{\rho_o}\ . \tag{3.6.8}$$

The last result follows from the use of the various thermodynamic relations given above (which also imply that $a^2_e = pC_{pe}/C_{ve}\rho$). Comparison of (3.6.8) with the viscous term $\frac{4}{3}\nu_o$ where ν_o is η_o/ρ_o shows that a rapidly-relaxing mode has an influence on the gas-dynamical (sound) waves that exactly parallels the effects of viscosity. It can be shown to operate only when there is dilatation in the medium ($\partial u_k/\partial x_k \neq 0$) and so η_v is called an <u>equivalent bulk viscosity</u>. <u>Rapidly</u>-relaxing internal modes can therefore have their influence summarised by replacing $\frac{2}{3}\eta$ in the coefficient of $\partial u_\ell/\partial x_\ell$ in (3.2.5) by

$(\frac{2}{3}\eta + \eta_v)$, replacing λ in (3.2.6) by $(\lambda + \rho\tilde{D}C)$ and $e_1(p,\rho,q)$ in (3.2.8) by $e_1(T_1)$.

Similar conclusions are reached by Wang Chang, et al, (1964) and Kogan (1969, §3.10) based on kinetic theory arguments that relate (some at least) of the occupation numbers n_{α} to local equilibrium values described by T_1. It is not trivial to note that the 'kinetic-theory' and the 'gas-dynamical' derivations, given here, are consistent!

4. ENERGY TRANSFER IN A 'REACTIVE' GAS

4.1 'Reaction' at a Surface

Assume that the 'reaction' term in (3.2.4) can be written in the form

$$w(p,T_1,q) = w_h(p,T_1,q) + w_w(p,T_1,q)\delta(y), \tag{4.1.1}$$

where $\delta(y)$ is a Dirac delta-function of $y \equiv x_2$, and it has been decided to express w as a function of p, T_1 and q rather than p, ρ and q. The quantity w_h is the 'reaction' term for the homogeneous gas phase, which is all that it has been necessary to deal with so far (e.g. in (3.3.2)). The new term in (4.1.1) that is proportional to w_w represents a heterogeneous 'reaction' that is localised in space and only occurs at a surface $y = 0$.

Assume for simplicity that all physical quantities depend only upon t and $x_2(\equiv y)$; combination of (3.2.4) with (4.1.1) gives

$$\frac{\partial}{\partial t}(\rho q) + \frac{\partial}{\partial y}(\rho v q) - \frac{\partial}{\partial y}(\rho\tilde{D}\,\frac{\partial q}{\partial y}) = w_h + w_w\delta(y), \tag{4.1.2}$$

where v is written for u_2. Integration of (4.1.2) with respect to y from $-\epsilon$ to ϵ, assuming that all of the physical quantities are continuous, followed by passage to the limit $\varepsilon \to 0$, gives

$$\rho_o q_o (v_+ - v_-) - \rho_o \tilde{D}_o \{ (\frac{\partial q}{\partial y})_+ - (\frac{\partial q}{\partial y})_- \} = w_w(p_o, T_o, q_o), \qquad (4.1.3)$$

where $(\ )_\pm$ implies evaluation at $y = 0\pm$.

The surface 'reaction' term w_w can be treated just like w in §3.3 (see (3.3.2) particularly), so that we can write

$$w_w \simeq -(\partial w_w/\partial q)_o (q_e - q). \qquad (4.1.4)$$

However $\partial w_w/\partial q$ is not related to a relaxation time, as in (3.3.3), in view of the different physical meanings of w_w and w_h. Evidently $-(\partial w_w/\partial q)$ must be positive and have the dimensions of density times velocity.

Since it is primarily the random thermal molecular motions that are responsible for bringing molecules into contact with a surface, and so permitting the energy transfer that is implicit in (4.1.2) to take place, it is clear that the velocity associated with $\partial w_w/\partial q$ must be of this same order; it will be assumed here to be equal to the appropriate frozen sound speed, a_{fo}. Thus $\partial w_w/\partial q$ is proportional to $\rho_o a_{fo}$; writing the constant of proportionality as $-\Gamma$ (evidently the value of Γ will depend on the nature of the gas/surface combination) a physically satisfactory version of (4.1.4) is

$$w_w = \Gamma a_{fo}(q_e - q). \qquad (4.1.5)$$

Now assume that the surface is an interface between the 'reactive' gas and a solid. There can then be no question of flow perpendicular to the surface, so that $v_\pm$ are both zero. In addition there is no question of diffusion of q within the solid, so that $(\partial q/\partial y)_\pm$ is zero depending on whether the gas is below or above the surface respectively.

Thus (4.1.3 and 5) give

$$\mp \tilde{D}_o \frac{\partial q}{\partial y} = \Gamma_o a_{fo}(q_e - q) \qquad (4.1.6)$$

as boundary conditions on the non-equilibrium variable q for gas above (below) a wall whose <u>catalytic efficiency</u> is Γ_o. It is clear that Γ_o can be reinterpreted in terms of <u>accommodation coefficients</u> in appropriate circumstances; in broad terms Γ in such circumstances is proportional to $r_q/(2 - r_q)$ where $r_q (0 < r_q < 1)$ is an accommodation coefficient for transfer of q at the surface; $q_e - q$ is then equivalent to an energy or temperature jump at the surface for the q-mode.

4.2 A Simple Heat-Conduction Cell

A reactive gas is contained between a pair of parallel flat plates situated at y = 0 and δ. With no flow and constant pressure between the plates, the presumed steady-state is governed by the energy and relaxation equations (3.2.3 and 4), which now read

$$-\lambda_o \frac{dT_1}{dy} - \rho_o \tilde{D}_o \frac{dq}{dy} = \dot{q}_w , \qquad (4.2.1)$$

$$-\tau_o \tilde{D}_o \frac{d^2 q}{dy^2} = q_e - q . \qquad (4.2.2)$$

It is assumed that differences in T_1 and q across the layer are small relative to the mean $(\ )_o$ values. In addition $0 < y < \delta$ so that only the homogeneous gas-phase reaction term w_h from (4.1.1) is needed in (4.2.2), and this source term has been expressed in the form given in (3.3.2). The constant q_w is the energy flux (per unit area per unit time) across the cell.

It is evidently sensible to use a dimensionless coordinate η, defined so that

$$\eta = y/\delta\,, \tag{4.2.3}$$

in place of y. Then (4.2.2) becomes

$$\Delta_o^2 \frac{d^2q}{d\eta^2} = q - q_e, \quad \Delta_o^2 \equiv \tau_o\tilde{D}_o/\delta^2\,, \tag{4.2.4a,b}$$

where Δ_o^2 is a dimensionless number (a Damköhler number) that is, in this case, a ratio of the relaxation time τ_o to a measure, $\delta^2/\tilde{D}_o$, of the time that it takes for q to diffuse across the cell.

If $\Delta_o \to 0$ (4.2.4a) shows that $q \to q_e$ and (4.2.1) then gives

$$q_w = -(\lambda_o + \rho_o\tilde{D}_o \frac{dq_e}{dT_1}) \frac{dT_1}{dy} \equiv -\lambda_{Eo} \frac{dT_1}{dy}\,, \tag{4.2.5}$$

since q_e depends only upon T_1 when p is fixed. In view of (3.6.5 and 7) it is obvious from (4.2.5) that energy is being conducted across the cell by the Eucken-corrected thermal conductivity λ_E under these essentially equilibrium conditions.

However, it must be remembered that not only must the translational temperature T_1 obey conditions

$$T_1 = T_{1\delta},\ T_{1w} \text{ at } y = \delta,\ 0, \tag{4.2.6}$$

but q must obey the accommodation conditions (4.1.6). These are re-written here as,

$$\mp \frac{dq}{d\eta} = \bar{\Gamma}_o(q_e - q), \ \eta = 0,1; \ \bar{\Gamma}_o \equiv a_{fo}\delta\Gamma_o/\tilde{D}_o\ . \tag{4.2.7}$$

There is no reason why Γ_o should be the same at $y = \delta$ as it is at $y = 0$ but it has been taken to be so here for simplicity.

Now (4.2.5) can only be correct if $dq/d\eta$ is the same as $CdT_1/d\eta$ (where C is dq_e/dT_1) everywhere, including as η approaches 0 and 1; but

$dT_1/d\eta$ cannot be zero, and yet (4.2.7) insists that it must be so at $\eta = 0,1$ when $q = q_e$; the only possibility for a consistent solution of this kind is to find $\bar{\Gamma}_o = \infty$, which it evidently is not! It is obviously the singular character of (4.2.4a) when Δ_o vanishes that is causing the trouble, and a full solution of this equation is necessary.

4.3 Hirschfelder Layers and Knudsen Layers

With small differences in T_1 and q across the cell, and the implicit limitation to relatively small differences between q and q_e, which can be written as $CT_1 - q$, the problem posed by (4.2.1 and 2) with conditions (4.2.6 and 7) is very simple in principle, but quite long-winded in practice! The results of the analysis are listed here:-

$$T_1 = -\frac{\dot{q}_w}{\lambda_{Eo}}\{y - \tfrac{1}{2}\delta - \delta(\frac{\lambda_{Eo}}{\lambda_o} - 1)\Delta_* F(y)\} + \tfrac{1}{2}(T_{1\delta} + T_{1w}), \quad (4.3.1)$$

$$q = C_o\{T_1 - \frac{\dot{q}_w\delta}{\lambda_o}\Delta_* F(y)\}, \quad (4.3.2)$$

$$-\dot{q}_w \frac{\delta}{\lambda_{Eo}}\{1 + 2(\frac{\lambda_{Eo}}{\lambda_o} - 1)\Delta_* F(o)\} = T_{1\delta} - T_{1w}, \quad (4.3.3)$$

where

$$\Delta_* = \Delta_o\sqrt{\lambda_o/\lambda_{Eo}}, \quad (4.3.4)$$

$$\tilde{a} = \Gamma_o a_{fo}\sqrt{\lambda_{Eo}\tau_o/\lambda_o\tilde{D}_o} = \bar{\Gamma}\lambda_{Eo}\Delta_*/\lambda_o, \quad (4.3.5)$$

$$F(y) = \{\exp([\delta - y]/\delta\Delta_*) - \exp(y/\Delta_*\delta)\}\{(\tilde{a} + 1)\exp(1/\Delta_*) - (\tilde{a} - 1)\}^{-1}. \quad (4.3.6)$$

Evidently Δ_* and Δ_o are of the same order of magnitude; since $\tilde{D}_o$ is roughly $a_{fo}^2\tau^{coll}$,

$$\Delta_* \sim \Delta_o \sim Kn\sqrt{\tau_o/\tau_o^{coll}}, \quad (4.3.7)$$

where the Knudsen number of the cell is defined here to be

$$Kn = a_{fo}\tau_o^{coll}/\delta \equiv \delta_{Kn}/\delta, \tag{4.3.8}$$

and must always be small for validity of the present theories. Thus although τ_o is usually much greater than τ_o^{coll}, Kn can be so small that $\Delta_* \ll 1$ and gas-phase equilibrium can be anticipated. When this is so it can be seen that both translational temperature T_1 and non-equilibrium variable q are effectively linear in y in the centre of the cell (where F(y) will be small; note that $F(\frac{1}{2}\delta)$ is always zero). Both T_1 and q will in general vary rather rapidly, and will be forced out of equilibrium, in layers adjacent to y = 0 and δ whose thicknesses are of order $\Delta_*\delta$, or δ_H (say).

Not only must these 'Hirschfelder layers', as they are sometimes called, be thin compared with the cell size when $\Delta_* \ll 1$, they must also be thick compared with the Knusden layers at y = 0 and δ. The latter are of thickness δ_{Kn} (see (4.3.8)), so that (4.3.7) makes

$$\delta_{Kn}/\delta_H \sim Kn/\Delta_* \sim \sqrt{\tau_o^{coll}/\tau_o} \ll 1. \tag{4.3.9}$$

Whilst (4.3.9) makes the point about the relative sizes of Knudsen and Hirschfelder layers, it is clearly important to note the square root factor in this relation.

Of course not only the existence but also the amplitude of departures from linearity of T_1 and q in the Hirschfelder layers depends upon the surface catalyticity, or accommodation factor, Γ_o, that appears in the quantity $\tilde{a}$ (see (4.3.5 and 6)). The order of magnitude of $\tilde{a}$ is evidently

$$\tilde{a} \sim \Gamma_o \sqrt{\tau_o/\tau_o^{coll}}\ . \tag{4.3.10}$$

Whilst the time-ratio is large, Γ_o may well be small in some circumstances and one cannot generalise about the size of $\tilde{a}$ except to say that the extreme value must be zero. When this happens $|F|$ is equal to one at both walls and the T_1 and q distortions in the Hirschfelder layers are then the largest possible.

4.4 Dominance of Surface Catalysis

The Hirschfelder-layer idea evidently fails when Δ_* is of order unity or greater. For very 'slow' modes, with τ_o substantially greater than τ_o^{coll}, $\Delta_* \gg 1$ can of course occur, even when Kn is suitably small for validity of continuum models. Taking the limit as $\Delta_* \to \infty$ (equivalent to letting $\tau_o \to \infty$) makes (cf (4.3.3))

$$\Delta_* F(0) \to 1/\{2 + (\lambda_{Eo}\bar{\Gamma}_o/\lambda_o)\} . \tag{4.4.1}$$

The { } term on the left-hand side of (4.3.3) behaves like

$$\{\ \} \to \left(\frac{\lambda_{Eo}}{\lambda_o}\right)\left[\frac{2 + \bar{\Gamma}_o}{2 + (\lambda_{Eo}\bar{\Gamma}_o/\lambda_o)}\right] \tag{4.4.2}$$

under these conditions, and it is clear that the heat-transfer rate $\dot{q}_w$ now depends critically on the catalyticity of the walls via $\bar{\Gamma}_o$.

If $\bar{\Gamma}_o$ vanishes { } is λ_{Eo}/λ_o and $\dot{q}_w$ depends upon λ_o; the internal mode is frozen in the homogeneous gas phase because τ_o is infinite, and is <u>not</u> excited by surface interaction because $\bar{\Gamma}_o$ is zero; as a result the mode does not participate in the energy-flux processes.

If, conversely, $\bar{\Gamma}_o$ itself is very large, { } $\to$ 1, and energy-flux takes place, in essence, via the conductivity λ_{Eo}. In view of the fact that homogeneous gas-phase 'reaction' is wholly inactive when Δ_* is infinite the foregoing result is rather remarkable, since it shows that efficient accommodation or catalysis of an internal mode can, by itself, establish a state of near-equilibrium between the internal and the translational modes.

In view of the fact that $\tilde{D}_o$ is roughly $a_{fo}^2 \tau^{coll}$ it can be seen, from (4.2.7) and the definition of Kn in (4.3.7), that

$$\bar{\Gamma}_o \sim \Gamma_o/Kn \;. \qquad (4.4.3)$$

Therefore large $\bar{\Gamma}_o$-values are quite probable, and existence of the accommodation-driven near-equilibrium conditions defined in the previous paragraph can be anticipated.

4.5 Comments on the Heat-Cell Solutions

There is clearly a good deal of interesting information contained in the simple heat-conductivity cell problem. Much of this information can be exploited in examinations of forced-convection problems, if only as an aid to the interpretation of results.

One large omission must be mentioned at this point. Although the foregoing work depends a lot on "q-jump" conditions at the walls, translational-temperature jumps are not even mentioned. The matter can be put right by relating $T_{1\delta,w}$ to the actual solid-surface temperatures in the usual way. Such a thing is out of place here (it can be studied in §§2.10 and 11 of the book by Clarke & McChsensy (1976) and in an article in Chapter 1 of the book edited by Wegener (1969), where additional references will be found). The present objective is to reveal purely polyatomic and 'reaction' effects, and this is most plainly done by keeping matters as simple as possible.

REFERENCES

1. Bird, R.B., Stewart, W.E. and Lightfoot, E.N. (1960) "Transport Phenomena". Wiley, New York.

2. Chapman, S. and Cowling, T.G. (1970) "The Mathematical Theory of Non-Uniform Gases", Third edition. Cambridge University Press.

3. Clarke, J.F. and McChesney, M. (1964) "The Dynamics of Real Gases". Butterworths, London.

4. Clarke, J.F. and McChesney, M. (1976) "Dynamics of Relaxing Gases". Butterworths, London.

5. Hirschfelder, J.O., Curtiss, C.F. and Bird, R.B. (1954) "The Molecular Theory of Gases and Liquids". Wiley, New York.

6. Jeffreys, H. (1952) "Cartesian Tensors". Cambridge University Press.

7. Kogan, M.H. (1969) "Rarefied Gas Dynamics". Plenum Press, New York.

8. Rich, J.W. and Treanor, C.E. (1970) "Vibrational Relaxation in Gas-Dynamics Flows". Ann Rev. Fluid Mech., 2, 355-396.

9. Stupochenko, Ye. Y, Losev, S.A. and Osipov, A.I. (1967) "Relaxation in Shock Waves". Springer-Verlag, Berlin.

10. Vincenti, W.G. and Kruger, C.H. (1965) "Physical Gas Dynamics". Wiley, New York.

11. Wang Chang, C.S., Uhlenbeck, G.E. and de Boer, J. (1964) "The Heat Conductivity and Viscosity of Polyatomic Gases". Studies in Statistical Mechanics, II, 243-268. North-Holland Pub. Co., Amsterdam.

12. Wegener, P.P. (Editor) (1969) "Nonequilibrium Flows", Part I; (1970) "Nonequilibrium Flows", Part II. Dekker, New York.

13. Williams, F.A. (1985) "Combustion Theory". Benjamin Cummings, Menlo Park, USA.

KINETIC EQUATIONS FROM HAMILTONIAN DYNAMICS: THE MARKOVIAN APPROXIMATIONS

H. Spohn
Universität München, München, F.R.G.

1. Introduction

The goal of non-equilibrium statistical mechanics is to explain the macroscopic behavior of matter from the dynamics of its microscopic constituents, i.e. atoms or molecules. Because of the large number of particles involved, this is a rather ambitious program. Therefore, as an intermediate step, one tries to write down an approximately valid dynamics as governed by a kinetic equation. Examples are the Boltzmann equation for a dilute gas, the hydrodynamic equations for an "aged" fluid, etc.. There is a lot known about the interrelationship between the microscopic and kinetic description of a physical system and it would be rather hopeless to press all these results into one lecture. Therefore I would like to do three things.

(i) The Lorentz gas and its various kinetic equations.
This is, presumably, the simplest, yet non-trivial, model. I am convinced that a full understanding of the Lorentz gas will help in more complicated situations.

(ii) The derivation of the Boltzmann equation for a hard sphere gas.
This is the work of Lanford [1,2], cf. also [3], and is, in my opinion one of the outstanding, more recent results, in the kinetic theory of gases.

(iii) Fluctuations for a dilute gas.
This is based on recent work of van Beijeren, Lanford, Lebowitz and myself [4].

Before entering to the Lorentz gas let me at least mention the general program for deriving a kinetic equation.

(I) The microscopic model.

A model is regarded as admissable, if it satisfies

(i) The dynamics of the model is governed by Hamilton's equation of motion (by the Schrödinger equation for a quantum mechanical system). Therefore the model is specified by a Hamiltonian, supplemented, if necessary, by proper boundary conditions.

(ii) Initially (at t = 0) a statistical state is given, i.e. a probability measure on the phase space associated with the classical system, resp. a statistical operator on the Hilbert space associated with the quantum mechanical system.

I emphasize that all statistical assumptions enter through the initial state.

To mention some of the models which are studied

	system + reservoir models	interacting particle systems
classical continuous models	Lorentz gas Rayleigh gas	classical gas classical fluid
classical lattice models	impurity in a lattice	harmonic, anharmonic lattice
quantum mechanical models	atom coupled to EM field electron in a solid	qm fluid, qm solid laser
	linear kinetic equation	non-linear kinetic equation

On second thought, one would like to understand why certain statistical assumptioms work so well or, more ambitiously, try to reduce or even to avoid statistical assumptions altogether. This is a rather difficult subject about which I have very little to say. I propose here to take the point of view of statistical mechanics and to regard the justification of the initial statistical ensemble as a separate problem.

(II) The approximation.

As already mentioned kinetic equations are good approximations only under certain physical conditions. This is translated into the theoretical framework by performing a certain limit. E.g. the Boltzmann equation is known to be valid for a dilute gas. Therfore in order to derive the Boltzmann equation one should let the density of the system go to zero, which is a condition on the initial state. However, then the mean free path and the mean free time, which are the typical length and time scale of the system, will tend to infinity. To obtain a well defined limit, therefore one has to adjust also the length and time scale appropriately.

It will turn out that the approximations can always be chosen in such a way that the Hamiltonian and the initial state are scaled. So, if we denote the scaling parameter by ε, then to each ε one has a well defined microscopic dynamics and, provided that the scaling is chosen appropriately, in the limit as $\varepsilon \to 0$ one obtains a limiting dynamics governed by a kinetic equation. In many cases to find the proper approximation is already a non-trivial problem.

(III) The proof of convergence to a limiting dynamics.

This is the more mathematical part of the program. Here, one has to prove a theorem which assures, under certain conditions on the scaled Hamiltonian and the scaled initial states, the convergence to a limiting dynamics.

The kinetic equations which I will discuss have a remarkable common feature: they are all first order in the time-derivative, i.e. they have no memory terms. The future state of the system is completely determined by its present state. This fact leads to a deeper probabilistic interpretation of the approximation leading to a kinetic equation. The limit can be understood as approximating a non-Markovian stochastic process by a Markov process.

2. The Lorentz Gas and its Various Markov Approximations

The Lorentz gas consists of a particle moving through infinitely heavy, randomly distributed scatterers. Let $x = (q,p) \in R^3 \times R^3$ denote the position and the momentum of the Lorentz particle. The mass of the Lorentz particle is set equal to one. We could restrict the motion of the Lorentz particle to some finite region, but it will be convenient not to do so. Then $Q = (q_1, q_2, \ldots)$ denotes a configuration of scatterers in R^3, where q_j is the center of the j-th scatterer. $(q_1, q_2, \ldots)$ is either a finite or a countable sequence in R^3. In every bounded region there should be only a finite number of scatterers. So Q is assumed to be locally finite. Let $\mathfrak{X}$ denote the space of all locally finite configurations. The interaction between the Lorentz particle and the scatterers is specified by a central potential V_ε of finite range. V_ε is assumed to be twice differentiable. ε is a scaling parameter which is introduced already here for convenience. Then the motion of the Lorentz particle is defined through the solution of Newton's equation of motion

$$\begin{aligned} \frac{d}{dt} q^\varepsilon(t,x,Q) &= p^\varepsilon(t,x,Q) \\ \frac{d}{dt} p^\varepsilon(t,x,Q) &= - \sum_j \nabla V_\varepsilon(q^\varepsilon(t,x,Q) - q_j) \, , \end{aligned} \tag{1}$$

$Q = (q_1, q_2, \ldots) \in \mathfrak{X}$, with initial conditions $q^\varepsilon(0,x,Q) = q$, $p^\varepsilon(0,x,Q) = p$. Since the scatterers are infinitely heavy, Q does not change in time. The sum in (1) makes sense, since V_ε is of finite range and since Q is locally finite. However, it may happen that the Lorentz particle reaches infinity in a finite time. For the distribution of scatterers to be considered below the set of such exceptional configurations is of measure zero. Although quite obvious, I emphasize that the joint system "Lorentz particle + scatterers" is of Hamiltonian form.

As initial distribution we choose $\delta_x \times \mu^\varepsilon$. δ_x is the point measure concentrated at $x \in R^6$. So the Lorentz particle starts at $x = (q,p)$. For simplicity μ^ε is choosen to be the ideal gas distribution with varying density, or equivalently to be the Poisson distribution. It is determined by the correlation functions

$$\rho_n^\varepsilon(q_1,\dots,q_n) = \prod_{j=1}^{n} \rho_1^\varepsilon(q_j) \quad . \tag{2}$$

$\rho_n^\varepsilon(q_1,\dots,q_n)$ is the expectation to find an n-tupel of scatterers at $q_1,\dots,q_n$.

In the physical literature it is customary to study the motion of the Lorentz particle through the <u>reduced</u> (or averaged) <u>dynamics</u>. The initial distribution of the Lorentz particle is assumed to be $f(x)dx$. Then for a fixed configuration Q the distribution of the Lorentz particle at time t is $f(x^\varepsilon(-t,x,Q))dx$, $x^\varepsilon(t,x,Q) = (q^\varepsilon(t,x,Q),p^\varepsilon(t,x,Q))$ and the averaged distribution of the Lorentz particle at time t is

$$f^\varepsilon(x,t)dx = (\int \mu^\varepsilon(dQ)\ f(x^\varepsilon(-t,x,Q))\ )dx. \tag{3}$$

The evolution $f(x) \to f^\varepsilon(x,t)$ defines the reduced dynamics. The evolution equation for $f^\varepsilon(x,t)$, which can be derived by the usual projection operator techniques, contains a memory kernel. In a Markovian approximation the potential V_ε and the distribution of scatterers μ^ε is scaled in such a way that the memory converges to a δ-function in time.

$\int_A dx f^\varepsilon(x,t)$ is the probability to find the Lorentz particle in the set $A \in R^6$ at time t given it started with the distribution $f(x)dx$. One may be interested in a more complicated information, such as the probability to find the Lorentz particle at time t_1 in $A_1,\dots,$ at time t_n in A_n given that it started initially at x. For this purpose it is then natural to regard the motion of the Lorentz particle as a <u>stochastic pro-</u>

cess. For a fixed configuration Q the position and the momentum of the Lorentz particle is determined through (1). But since the distribution of the scatterers is random, the position and the momentum of the Lorentz particle at time t is also random. We denote these random variables by $X^{\varepsilon}(t)$,

$$X^{\varepsilon}(t): Q \to q^{\varepsilon}(t,x,Q), p^{\varepsilon}(t,x,Q) \ . \tag{4}$$

The above mentioned probability is then

$$P_x^{\varepsilon}(X^{\varepsilon}(t_1) \in A_1, \ldots, X^{\varepsilon}(t_n) \in A_n) \ . \tag{5}$$

P_x^{ε} denotes the probability measure, where the subscript x indicates that the particle starts at x. P_x^{ε} weighs the different possible histories of the Lorentz particle.

A Markovian approximation means that the non-Markovian process $X^{\varepsilon}(t)$ converges as $\varepsilon \to 0$ to a Markov process. Technically, the measure P_x^{ε} converges weakly to a measure P_x which determines a Markov process. Suppose that the limiting Markov process is homogeneous in time and that it is given through the transition probability $p_t(x'|x)dx'$. ($\int_A dx' p_t(x'|x)$ is the probability to find the particle in A at time t given that it started at x initially.) Then, up to some technical points, the convergence to the limiting Markov process is defined by

$$\lim_{\varepsilon \to 0} P_x^{\varepsilon}(X^{\varepsilon}(t_1) \in A_1, \ldots, X^{\varepsilon}(t_n) \in A_n)$$
$$= \int_{A_n} dx_n \ldots \int_{A_1} dx_1 \, p_{t_n - t_{n-1}}(x_n|x_{n-1}) \ldots p_{t_1}(x_1|x) \ . \tag{6}$$

for all measurable sets $A_1, \ldots, A_n$, and all $0 < t_1 < \ldots < t_n$. In (6) we used the Chapman-Kolmogorov equations.

A probability density for the limiting Markov process evolves as $f(x,t) = \int dx' p_t(x|x') f(x')$. Differentiating with to t, using again the Chapman-Kolmogorc equations,

$$\frac{\partial}{\partial t} f(x,t) = Lf(x,t) \tag{7}$$

with some linear operatorLacting on f. (7) will be the kinetic equation we wanted to derive. The transition probability $p_t(x'|x)$ is the kernel of e^{Lt}.

Thus the goal is to show that the stochastic dynamics of the Lorentz particle converges in a suitable limit to the stochastic dynamics as determined through the kinetic equation.

2.1 The Low Density Limit (Grad Limit)

We let the density of scatterers tend to zero proportional to ε. Then the typical spacial scale for the Lorentz particle is on the order of a mean free path, i.e. on the order ε^{-1}, and the typical time scale is on the order of a mean free time, i.e. on the order ε^{-1}. To obtain a non-trivial limit we therefore have to scale also space and time,

$$q_\varepsilon = \varepsilon^{-1} q\ , \quad t_\varepsilon = \varepsilon^{-1} q\ . \tag{8}$$

In order that the Lorentz particle still sees the spacial variation,

$$\rho_1^\varepsilon(q_\varepsilon) = \varepsilon r(\varepsilon q_\varepsilon)\ . \tag{9}$$

It is convenient to go over to the t, q scale. Then, equivalently

q,p,t are unscaled

$$\rho_1^\varepsilon(q) = \varepsilon^{-2} r(q) \tag{10}$$

(ε^{-d+1} in d dimensions) and

$$V_\varepsilon(q) = V(q/\varepsilon) \ , \tag{11}$$

where V is a given central, twice differentiable potential of finite range. Note that the normalized differential cross section $\sigma(dp'|p)$ remains unchanged under the scaling (11). $\sigma(dp'|p)$ is the probability that the particle is scattered into dp' given its incoming momentum p.) As intended the mean free path of the Lorentz particle remains constant under the scaling (10) and (11). Typically, the Lorentz particle travels freely a mean free path, is then scattered according to the differential cross section, travels freely a mean free path, etc.. Since the the radius of the scatterers shrinks as ε the probability of recollisions goes to zero.

Let $X^\varepsilon(t)$ be the stochastic motion of the Lorentz particle scaled according to (10) and (11). Then one proves [5] that, if r is bounded and continuous, $X^\varepsilon(t)$ converges to $X(t)$ as $\varepsilon \to 0$, where $X(t)$ is the Markov process governed by the linear Boltzmann equation (Rayleigh-Boltzmann equation)

$$\frac{\partial}{\partial t} f(q,p,t) = -p\cdot\nabla_q f(q,p,t) + r(q)\pi|p|\int\sigma(dp'|p)\{f(q,p',t) - f(q,p,t)\} \tag{12}$$

$\sigma(dp'|p)$ contains the δ-function $\delta(|p'| - |p|)$ because of conservation of energy in a collision.

(12) is the forward equation of the following Markov process: The particle starts at q with momentum p. $p(t) = p$ for $0 \leq t < t_1$, where t_1 is a random time with distribution $\exp\{-\pi|p|\int_0^{t_1} ds\ r(q+ps)\}$. Then, independently of t_1, p jumps from p to p_1 with probability $\sigma(dp_1|p)$. $p(t) = p_1$ for $t_1 \leq t < t_1 + t_2$, where t_2 is a random time with distribution $\exp\{-\pi|p|\int_0^{t_2} ds\ r(q+pt_1+p_1 s)\}$, etc..

2.2 The Weak Coupling Limit

We consider a situation, where the scatterer potential is weak,

$$V_\varepsilon(q) = \varepsilon^{1/2} V(q) \quad . \tag{13}$$

Then the change of momentum in a collision is of the order $\varepsilon^{1/2}$. A non-trivial limit is obtained provided the Lorentz particle suffers ε^{-1} collisions per unit time. For constant density of scatterers this is achieved by sacling time as

$$t_\varepsilon = \varepsilon^{-1} t \ . \tag{14}$$

In order not to loose sight of the Lorentz particle also space has to be scaled as

$$q_\varepsilon = \varepsilon^{-1} q \ . \tag{15}$$

(The free motion remains invariant under (14) and (15).) Again it is convenient to go over to the t,q scale. Then, equivalently,

q,p,t remain unscaled,

$$\rho_1^\varepsilon(q) = \varepsilon^{-3} r(q) \tag{16}$$

(ε^{-d} in d dimensions) and

$$V_\varepsilon(q) = \varepsilon^{1/2} V(q/\varepsilon) \quad . \tag{17}$$

Note that a constant fraction of space remains filled by scatterers.

Typically the Lorentz particle travels freely a distance ε. It is then scattered over a time span ε resulting in a momentum change of the order $\varepsilon^{1/2}$, etc.. Therefore one expects that in the limit $p(t)$ diffuses constained by $|p(t)| = |p|$. Indeed by second order pertubation theory (or, if so wanted, by projection operator techniques) one formally obtains in the limit $\varepsilon \to 0$ as evolution equation for the Lorentz particle the linear Landau equation

$$\frac{\partial}{\partial t} f(q,p,t) = - p \cdot \nabla_q f(q,p,t) + \alpha r(q) \left[\sum_{i,j=1}^{3} \frac{\partial}{\partial p_i} D_{ij}(p) \frac{\partial}{\partial p_j} \right] f(q,p,t) \tag{18}$$

where $p = (p_1, p_2, p_3)$ and

$$D_{ij}(p) = \frac{1}{2|p|^3} (|p|^2 \delta_{ij} - p_i p_j) \quad ,$$

$$\alpha = \frac{\pi}{2} \int dk |k| |\hat{V}(k)|^2 \tag{19}$$

with $\hat{V}$ the Fourier transform of V. The term in the square brackets is just $(1/|p|) \times$Laplace-Beltrami operator on a sphere with radius $|p|$ as anticipated.

Let $X^\varepsilon(t)$ be the stochastic motion of the Lorentz particle scaled according to (16) and (17). Papanicalaou [6] has shown recently that, if V is of finite range and if r is constant and small enough, then $X^\varepsilon(t)$ converges to a Markov process $X(t)$ as $\varepsilon \to 0$, where $X(t)$ is the diffusion process governed by (18) as forward equation.

2.3 The Mean Field Limit

The spirit of a mean field approximation is that the force between any pair of particles goes to zero, whereas the mean force exerted on a particular particle by all the other particles remains finite. This is achieved by scaling

$$V_\varepsilon(q) = \varepsilon V(q) \tag{20}$$

and

$$\rho_1^\varepsilon(q) = \varepsilon^{-1} r(q) . \tag{21}$$

If we consider a scale on which the density of scatterers is constant, then, equivalently,

$$q_\varepsilon = \varepsilon^{-1/3} q \ , \quad t_\varepsilon = \varepsilon^{-1/3} t \tag{22}$$

and

$$V_\varepsilon(q) = V(\varepsilon^{1/3} q) . \tag{23}$$

Therefore, physically, the mean field limit corresponds to a situation with a weak, long range scatterer potential.

Let $F_\varepsilon(q,Q) = \varepsilon \sum_j F(q - q_j)$, $Q = (q_1, q_2, \ldots)$, where $F = -\nabla V$ is the force. Then

$$\begin{aligned} &\int \mu^\varepsilon(dQ) \, |F_\varepsilon(q,Q) - \int dq' r(q') F(q - q')|^2 \\ &= \varepsilon \int dq_1 r(q_1) \, |F(q - q_1)|^2 . \end{aligned} \tag{24}$$

Therefore in the limit $\varepsilon \to 0$ $F_\varepsilon(q,\cdot)$ converges in probability to the mean force $\overline{F}(q) = \int dq' r(q') F(q - q')$, i.e. for small ε most configurations yield almost the mean force on the Lorentz particle. One expects then that the motion of the Lorentz particle becomes deterministic in the limit $\varepsilon \to 0$ and is governed by the effective force $\overline{F}$.

Indeed, using the methods of [7,8], one proves that the stochastic motion $X^\varepsilon(t)$ of the Lorentz particle, scaled according to (20) and (21), converges as $\varepsilon \to 0$ to the deterministic motion governed by

$$\frac{d}{dt} q(t) = p(t) \ , \ \frac{d}{dt} p(t) = \overline{F}(q(t)) \tag{25}$$

with initial condition q,p, provided that the potential V is twice differentiable with bounded derivatives and that r is bounded.

In analogy to the two cases before, (25) may be called the linear Vlasov equation.

2.4 The Hydrodynamic Limits

So far we obtained a Markovian approximation by letting a physical parameter, namely either the density orthe interaction strength or the inverse range of the potential, of the Lorentz gas tend to zero. Physically, one would like to know properties of the Lorentz gas at finite interaction strength and constant density. Amazingly enough, in this situation there is yet another Markovian approximation possible, although of a somewhat more subtle nature. For a fluid the same approximations lead to the hydrodynamic equations. Therefore we call these approximations hydrodynamic limits.

The hydrodynamic limits are discussed on a much deeper level by Prof. Sinai in his lecture. Nvertheless I would like to include here some sketchy remarks, since to my experience the difference between the hydrodynamic limits and the limits discussed previously are often not clearly understood.

The physical intuition behind the hyrdynamic limit is that the system reaches relatively fast local equilibrium characterized by the hydrodynamic fields corresponding to locally conserved quantities and that subsequently these fields evolve relatively slowly. Therefore the hydrodynamic limits involve the long time behavior of the system and this is the reasonwhy they are degrees more difficult to prove than the limits encountered so far.

For the Lorentz gas only the number of particles is conserved. Therfore the only hydrodynamic field is the spacial density $\rho(q,t)$. The analogue of the Euler equations is simply

$$\frac{\partial}{\partial t} \rho(q,t) = 0 \tag{26}$$

and the analogue of the Navier Stokes equations is the diffusion equation

$$\frac{\partial}{\partial t} \rho(q,t) = D \, \Delta_q \, \rho(q,t) \quad . \tag{27}$$

Let $q(t), p(t)$ be the stochastic motion of the Lorentz particle for fixed interaction potential and for constant density of scatterers. For the hydrodynamic limits one considers only the spacial part of the stochastic motion

$$q(t) = q + \int_0^t ds \; p(s) \quad . \tag{28}$$

To obtain the analogue of the Euler equations one has to consider the evolution of the Lorentz particle over distances on the order ε^{-1} and over times on the order ε^{-1}. This leads to the scaling

$$q^\varepsilon(t) = q + \varepsilon \int_0^{\varepsilon^{-1}t} ds \; p(s) \quad . \tag{29}$$

One should show that $q^\varepsilon(t) \to q$ as $\varepsilon \to 0$. This seems to be reasonable, since one expects the Lorentz particle to be a distance ε^{-1} away from q in a time ε^{-2}. Note that one has to prove essentially the ergodicity of the momentum process. For certain fixed periodic configurations of hard disk scatterers this follows from Sinai's beautiful results about the billiard, cf. Gallavotti's lectures [9]. (Sinai proves, in fact, much stronger properties than ergodicity.)

To obtain the diffusion equation one has to consider the evolution of the Lorentz gas over distances on the order ε^{-1} and over times on the order ε^{-2}. This leads to the scaling

$$q^\varepsilon(t) = q + \varepsilon \int_0^{\varepsilon^{-2}t} ds \; p(s) \quad . \tag{30}$$

One should show that $q^{\varepsilon}(t) \to q^{0}(t)$ as $\varepsilon \to 0$, where $q^{0}(t)$ is the stochastic process corresponding to the diffusion equation (27), i.e. $q^{0}(t)$ is the Wiener process starting at q with diffusion constant 2D.

(30) implies the validity of the diffusion equation in the following sense: If $\rho(q,t)dq$ denotes the (averaged) spacial distribution of the Lorentz particle at time t given that it started with the distribution $f(q,p)dqdp$ initially, then $\rho^{\varepsilon}(q,t) = \varepsilon^{-3}\rho(\varepsilon^{-1}q,\varepsilon^{-2}t)$ converges weakly to $(\pi 4Dt)^{-3/2}\ \exp\{-q^2/4Dt\}$.

Of course the first thing to show would be the existence of a finite diffusion constant D, which means to prove

$$\lim_{t\to\infty} \frac{1}{t}\, E((q(t) - q)^2) = 2dD\ , \tag{31}$$

d = dimension. The mean square displacement should be proportional to t for large times. Even the existence of a finite diffusion constant for the Lorentz gas is an open, presumably very difficult problem. (Sinai proves (30), in particular (31), for certain fixed periodic configurations of hard disk scatterers.)

I did not spell out the complete picture: For a given potential, as one increases the density of the scatterers at some stage the Lorentz particle may be trapped. The motion of the Lorentz particle changes then qualitatively in so far as the mean square displacement grows more slowly than t.

We are now in a position to understand somewhat better the approximations considered before. For a given potential let us denote the coupling strength by λ and the constant density of scatterers by ρ. We indicate the dependence of the diffusion constant D on λ and ρ by $D(\lambda,\rho)$. We consider a spacial scale on which the range of the potential remains constant. Then as either $\lambda \to 0$ or $\rho \to 0$ $D(\lambda,\rho) \to \infty$, i.e. the particle diffuses more and more freely. However, when discussing the low density and weak coupling limit, we scaled also the range of the potential or, equivalently, the spacial scale. Using the definition (31) of $D(\lambda,\rho)$ formally the scaling is such that

$$\begin{aligned} \rho D(\lambda,\rho) &\rightarrow D_B(\lambda) \quad \text{for } \rho \rightarrow 0 , \\ \lambda^{1/2} D(\lambda,\rho) &\rightarrow D_L(\rho) \quad \text{for } \lambda \rightarrow 0 , \end{aligned} \tag{32}$$

where $D_B(\lambda)$ is the diffusion constant obtained from the linear Boltzmann equation (12) and where $D_L(\lambda)$ is the diffusion constant obtained from the linear Landau equation (18). In fact, it is known [10] that scaling the Markov processes corresponding to (12) and to (18) as in (30) leads to a Wiener process with diffusion constant $2D_B(\lambda)$, resp. $2D_L(\rho)$. This means that the linear Boltzmann equation and the linear Landau equation predict qualitatively correctly the long time behavior of the Lorentz gas, although with the wrong diffusion constant.

Let me, finally, draw your attention to a very nice survey article on the Lorentz gas by Hauge [11], where a number of other topics are discussed, and on the detailed numerical studies by Lewis and Tjon [13] and by Alder and Alley [12] with references to older work.

3. The Derivation of the Boltzmann Equation

I now turn to interacting systems. The situation becomes more complex, but the basic strategy as developed for the Lorentz gas carries over unchanged. The low density limit for a system of hard spheres has been proved by Lanford [1,2] and for positive potentials with certain regularity conditions by King [3]. The main draw back at present is that the convergence can be controlled only over a time span on the order of a mean free time. The low density limit leads to the Boltzmann equation. I will describe Lanford's result below. The mean field limit has been proved by Braun and Hepp [8] and independently by Neunzert [7] for a general class of potentials and for all times. The mean field limit leads to the Vlasov equation. The weak coupling limit, which would yield the Landau equation, is unproved. Hydrodynamic limits look hopelessly difficult.

For the Boltzmann equation we use, for technical reasons, a somewhat different setting than before. Generalizations are possible.

With the same reasoning as for the Lorentz gas we immediately adopt the scaling (10) and (11), i.e. the density of particles increaseases as ε^{-2}, whereas the range of the potential decreases as ε.

We consider a bounded region $\Lambda \times R^3$ with smooth boundaries $\partial\Lambda$. Inside Λ we have hard spheres of mass one and diameter ε. (This corresponds to scaling the potential as in (11).) The number of particles is not necessarily fixed. So the classical phase space is $\Gamma = \bigcup_{n\geq 0} (\Lambda\times R^3)^n$. (Not all points of the phase space can be realized because of the hard core exclusion.)

The spheres (particles) are elastically reflected amongst themselves and at the boundary $\partial\Lambda$. For pair collisions and collisions with the wall the dynamics is thereby well defined. For grazing collisions and triple and higher collisions the dynamics simply remains undefined. Alexander [14] has shown that the complicated set of initial phase points which lead to such higher collisions at any previous or later time is of Lebesgue measure zero. Therefore, if the initial distribution of hard spheres is absolutely continuous with respect to Lebesgue, such exceptional initial phase points are of probability zero and can discarded.

Let the initial state μ^ε of the system be specified by the absolutely continuous probabilities of finding exactly n particles at $dx_1 \ldots dx_n$, $\{f_n^\varepsilon(x_1,\ldots,x_n)\frac{1}{n!}dx_1\ldots dx_n \mid n \geq 0\}$. Here $x_i = (q_i,p_i) \in \Lambda\times R^3$ stands for the position and the momentum of the i-th particle. I introduced the scaling parameter ε already here for convenience. Then the correlation functions $\{\rho_n^\varepsilon \mid n \geq 0\}$ corresponding to this state are defined by

$$\rho_n^\varepsilon(x_1,\ldots,x_n) = \sum_{m=0}^{\infty} \frac{1}{m!} \int_{(\Lambda\times R^3)^m} dy_1\ldots dy_m f_{n+m}^\varepsilon(x_1,\ldots,x_n,y_1,\ldots,y_m) \tag{33}$$

The time evolution of a state of the hard sphere system is studied by means of the time evolution of the corresponding correlation functions. A straightforward computation which is however non-trivial to justify rigorously [15,16] leads to the following evolution equation

$$\frac{\partial}{\partial t} \rho_n^\varepsilon(x_1,\ldots,x_n,t) = H_n^\varepsilon \rho_n^\varepsilon(x_1,\ldots,x_n,t) \tag{34}$$
$$+ \varepsilon^2 \sum_{j=1}^{n} \int_{\Lambda\times R^3} dp_{n+1} \int_{S^2} d\omega\omega\cdot(p_{n+1}-p_j)\rho_{n+1}^\varepsilon(x_1,\ldots,x_n,q_j+\varepsilon\omega,p_{n+1},t) .$$

Here ω is a unit vector in R^3 and $d\omega$ the surface measure of the unit sphere S^2 in three dimensions. H_n^ε describes the evolution of n hard spheres of diameter ε inside Λ. (34) is the <u>BBGKY hierarchy</u> for hard spheres.

We want to study the low density (Grad) limit of the solutions of the BBGKY hierarchy. As for the Lorentz gas the density is increased as ε^{-2}. (On this scale the mean free path remains constant, whereas the volume occupied by spheres tends to zero as $\varepsilon \to 0$.) Therefore for each hard sphere diameter ε one chooses an initial state with correlation functions ρ_n^ε such that $\rho_n^\varepsilon \sim \varepsilon^{-2n}$. With this in mind we define the <u>rescaled correlation functions</u>

$$r_n^\varepsilon(x_1,\ldots,x_n) = \varepsilon^{2n}\rho_n^\varepsilon(x_1,\ldots,x_n) . \tag{35}$$

Then (34) reads

$$\frac{d}{dt} r_n^\varepsilon(t) = H_n^\varepsilon r_n^\varepsilon(t) + C_{n,n+1}^\varepsilon r_{n+1}^\varepsilon(t) , \tag{36}$$

where the collision term is abbreviated as $C_{n,n+1}^\varepsilon$. Regarding the sequence $\{r_n^\varepsilon \mid n \geq 0\}$ as vector r^ε one can write (36) more compactly as

$$\frac{d}{dt} r^{\varepsilon}(t) = H^{\varepsilon}r^{\varepsilon}(t) + C^{\varepsilon}r^{\varepsilon}(t) \quad , \tag{37}$$

where H^{ε} is a diagonal matrix with entries H_n^{ε} and C^{ε} is a matrix with entries $C_{n,n+1}^{\varepsilon}$ and zero otherwise.

Eventually, we are interested in the time evolution of the rescaled one-particle correlation function, which gives the scaled average number of particles in an arbitrary region $\Delta \subset \Lambda \times R^3$ at time t. Therefore we adopt here the point of view of what was called reduced dynamics for the Lorentz gas. There is a way [5] to introduce a stochastic dynamics which turned out to be so natural in the case of the Lorentz gas. This leads the to the notion of a non-linear Markov process. (This is a Markov process, where the transition probability per unit time depends on the state of the system at the same time.) The relation between the average number and actual number of particles in a region Δ at time t will be discussed in the chapter about fluctuations.

Let me now consider H^{ε} as the unperturbed part of the operator $H^{\varepsilon} + C^{\varepsilon}$ and C^{ε} as the pertubation. The time-dependent (Dyson) pertubation series for the solution of (37) then reads

$$r^{\varepsilon}(t) = \sum_{m=0}^{\infty} \int_{0 \leqslant t_1 \leqslant \ldots \leqslant t_m \leqslant t} dt_m \ldots dt_1 \;\; S^{\varepsilon}(t-t_m)C^{\varepsilon}\ldots C^{\varepsilon}S^{\varepsilon}(t_1)r^{\varepsilon} \quad , \tag{38}$$

where r^{ε} stands for $r^{\varepsilon}(0)$, and where $(S^{\varepsilon}(t)r^{\varepsilon})_n = (\exp\{H^{\varepsilon}t\}r^{\varepsilon})_n$ $= \exp\{H_n^{\varepsilon}t\}r_n^{\varepsilon}$ gives the evolution of n hard spheres of diameter ε inside Λ, including the specular reflection at $\partial\Lambda$. Solutions of the BBGKY hierarchy are always understood in the sense of (38). Of course, we will have to say in what sense (38) converges.

To see what the formal limit $\varepsilon \to 0$ of (38) should be, let me consider the typical term

$$S^{\varepsilon}(s)\; C_{n,n+1}^{\varepsilon}\; S^{\varepsilon}(t)\; r_{n+1}^{\varepsilon} \tag{39}$$

with t,s > 0. (39) describes the evolution of n hard spheres for a time s, adjoining the (n+1)-st sphere with momentum p_{n+1} and position such as to touch the j-th sphere, the evolution of n+1 spheres for a time t, and, finally, the sum (integral) over all j = 1,...,n, momenta p_{n+1} and possible touching points $\varepsilon\omega$. Besides the possible collisions at time s at the point, where the j-th and the (n+1)-st sphere are joined, other collisions may occur. These are, however, of zero pobability in the limit $\varepsilon \to 0$ and (39) converges to

$$S(s)\, C_{n,n+1}\, S(t)\, r_{n+1} \, , \tag{40}$$

where S(t) is the free motion including specular reflection at the boundary and

$$\begin{aligned}(C_{n,n+1}\, r_{n+1})(x_1,\ldots,x_n) = \sum_{j=1}^{n} \int_{\omega\cdot(p_j-p_{n+1})\geq 0} dp_{n+1} d\omega \;\; \omega\cdot(p_j - p_{n+1}) \\ \{r_{n+1}(x_1,\ldots,q_j,p'_j,\ldots,q_j,p'_{n+1}) \\ - r_{n+1}(x_1,\ldots q_j,p_j,\ldots,q_j,p_{n+1})\} \, .\end{aligned} \tag{41}$$

p'_j, p'_{n+1} are outgoing momenta to the incoming momenta p_j, p_{n+1} and momentum transfer in direction ω. Therefore, formally as $\varepsilon \to 0$ (38) goes over to

$$r(t) = \sum_{m=0}^{\infty} \int_{0\leq t_1<\ldots\leq t_m\leq t} dt_m \ldots dt_1 \;\; S(t-t_m)C\ldots CS(t_1)r \, . \tag{42}$$

Differentiating (42) with respect to t one obtains the Boltzmann hierarchy for hard spheres

$$\begin{aligned}\frac{\partial}{\partial t} r_n(x_1,\ldots,x_n,t) = -\sum_{j=1}^{n} p_j\cdot\nabla_{q_j} r_n(x_1,\ldots,x_n,t) \\ + (C_{n,n+1}\, r_{n+1})(x_1,\ldots,x_n,t) \, .\end{aligned} \tag{43}$$

$-p_j\cdot\nabla_{q_j}$ includes the specular reflection at $\partial\Lambda$.

For $t < 0$ the same reasoning leads again to (43), but with the sign of the collision term reversed.

To prove that $r^{\varepsilon}(t)$ defined by (38) indeed converges to $r(t)$ defined by (42) as $\varepsilon \to 0$ we need two conditions.

Firstly, theinitial correlation functions r^{ε} have to be bounded uniformly in ε. This guarantees the uniform convergence of the pertubation series (38) for some time interval $|t| < t_o$. (Heuristicly, the finite radius of convergence comes from the following fact: For $n = 1$, in (38) the time integration yields $t^m/m!$, whereas the m collision operators yield $m!$. For a better result cancellations have to be taken into account.) If h_β denotes the normalized Maxwellian at inverse temperatre β, then a suitable bound is

(C1) There exist a pair (z,β) such that

$$\varepsilon^{2n}|\rho_n^{\varepsilon}(x_1,\ldots,x_n)| = |r_n^{\varepsilon}(x_1,\ldots,x_n)| \leqslant M\, z^n \prod_{j=1}^{n} h_\beta(x_j) \qquad (44)$$

for all $\varepsilon < \varepsilon_o$ with a positive constant M independent of ε.

Secondly, r_n has to converge to r_n in such a way that the series (38) converges term by term to the series (42). For the initial phase point $x^{(n)} = (x_1,\ldots,x_n) \in (\Lambda\times R^3)^n$ let $q_j(t,x^{(n)})$, $j = 1,\ldots,n$, be the position of the j-th point particle at time t under free motion and specular reflection at $\partial\Lambda$. Then

$$\Gamma_n(t) = \{x^{(n)} = (x_1,\ldots,x_n) \in (\Lambda\times R^3)^n \mid q_i(s,x^{(n)}) \neq q_j(s,x^{(n)})$$

$$\text{for } i \neq j = 1,\ldots,n \text{ and } -t \leqslant s \leqslant 0, \text{ if } t \geqslant 0,\ 0 \leqslant s \leqslant -t, \text{ if } t \leqslant 0\}$$

In words, $\Gamma_n(t)$ is the restriction of the n-particle phase space to the set of all phase points that under free backward streaming over a time t, if t is positive, (or free forward streaming over a time $|t|$, if t is negative) do not lead to a collision between any pair of particles, regarded as point particles. By this restriction only a seteof Lebesgue measure zero is excluded from $(\Lambda\times R^3)^n$.

Note that (i) $\Gamma_n(t)$ depends only on the free motion, (ii) $\Gamma_n(t) \subset \Gamma_n(t')$ for $t' = \alpha t$, $\alpha \geq 1$, (iii) $\Gamma_n(t) \neq \Gamma_n(-t)$ and (iv) that $x^{(n)} \in \Gamma_n(t)$ is equivalent to $\bar{x}^{(n)} \in \Gamma_n(-t)$, where $\bar{x}^{(n)}$ is the phase point obtained from $x^{(n)}$ under the reversal p_j to $-p_j$. In particular $\Gamma_n(t)$ is not invariant under reversal of velocities.

The suitable choice of convergence is then

(C2) There exists a continuous function r_n on $(\Lambda\times R^3)^n$ such that

$$\lim_{\varepsilon\to 0} \varepsilon^{2n} \rho_n^\varepsilon = \lim_{\varepsilon\to 0} r_n^\varepsilon = r_n \tag{45}$$

uniformly on compact sets of $\Gamma_n(s)$ for some $s \geq 0$.

Theorem (Lanford). Let$\{\rho_n^\varepsilon | n \geq 0\}$ be a sequence of initial correlation functions (not necessarily coming from a positive measure) of a fluid of hard spheres of diameter ε inside the bounded region Λ and let $\{\rho_n^\varepsilon | n \geq 0\}$ satisfy (C1) and (C2). Let $\rho_n^\varepsilon(t)$ be the solution of the BBGKY hierarchy with initial conditions ρ_n^ε and let $r_n(t)$ be the solution of the Boltzmann hierarchy with initial conditions r_n.

Then there exists a $t_o(z,\beta) > 0$ such that for $0 \leq t \leq t_o(z,\beta)$ the series (38) and (42) converge pointwise and such that $\rho_n^\varepsilon(t)$ satisfies a bound of the form (C1) with $z' > z$ and $\beta' < \beta$. Furthermore

$$\lim_{\varepsilon\to 0} \varepsilon^{2n}\rho_n^\varepsilon(t) = \lim_{\varepsilon\to 0} r_n^\varepsilon(t) = r_n(t) \tag{46}$$

uniformly on compact sets of $\Gamma_n(t+s)$.

For $-t_o(z,\beta) \leq t \leq 0$ (46) holds provided that $s \leq 0$ and that in the Boltzmann hierarchy the collision term $C_{n,n+1}$ is replaced by $-C_{n,n+1}$.

$t_o(z,\beta)$ may be choosen as $(1/5)(\sqrt{3}/\pi z\sqrt{\beta})$. The second factor has physically the meaning of the mean free time of a Boltzmann gas in equilibrium at inverse temperature β and density z.

The reader my wonder how Lanford's theorem escapes the conflict between the reversible character of the BBGKY hierarchy and the irreversible character of the Boltzmann hierarchy. The point is that if for the sequence of states $\rho^{\varepsilon}(t)$, $0 < t < \frac{1}{2}t_o(z,\beta)$, we reverse all momenta at time t and thereby form a new sequence of initial states $\tilde{\rho}^{\varepsilon}$, then $\tilde{\rho}^{\varepsilon}$ does not satisfy the condition (C2) of Lanford's theorem and therefore the theorem is not (and better should not be) applicable to this sequence of initial states.

The Boltzmann hierarchy has the well known property of "propagation of molecular chaos": If the initial conditions of the Boltzmann hierarchy factorize,

$$r_n(x_1,\ldots,x_n) = \prod_{j=1}^{n} r(x_j) \quad , \tag{47}$$

then the solutions with this initial condition stay factorized,

$$r_n(x_1,\ldots,x_n,t) = \prod_{j=1}^{n} r(x_j,t) \quad . \tag{48}$$

$r(x,t)$ is the solution of the <u>Boltzmann equation</u>

$$\frac{\partial}{\partial t} r(q,p,t) = -p\cdot\nabla_q r(q,p,t) + \int_{\omega\cdot(p-p_1)\geq 0} dp_1\, d\omega\, \omega\cdot(p-p_1) \{r(q,p',t)r(q,p_1',t) - r(q,p,t)r(q,p_1,t)\} \tag{49}$$

with initial conditions $r(q,p)$.

Lanford's theorem together with molecular chaos implies the validity of the Boltzmann equation in the following sense. Assume that the initial correlation functions of the hard sphere system $\{\rho_n^{\varepsilon} | n \geq 0\}$ satisfy the bound (C1) and that

$$\lim_{\varepsilon\to 0} \varepsilon^{2n} \rho_n^\varepsilon(x_1,\dots,x_n) = \prod_{j=1}^{n} r(x_j) \tag{50}$$

uniformly on compact sets of $\{x_1,\dots,x_n \in (\Lambda\times R^3)^n \mid q_i \neq q_j, i \neq j, i,j = 1,\dots,n\}$ with some continuous r. Then for $0 \leqslant t \leqslant t_o(z,\beta)$

$$\lim_{\varepsilon\to 0} \varepsilon^2 \rho_1^\varepsilon(x_1,t) = r(x_1,t) \tag{51}$$

uniformly on compact sets of $\Lambda\times R^3$, where $r(x_1,t)$ is the solution of the Boltzmann equation (49) with initial condition $r(x_1)$.

4. Fluctuations

Let $\Delta \subset \Lambda\times R^3$ be a bounded region. Then the average number of particles in Δ at time t is $\int_\Delta dx_1 \rho_1^\varepsilon(x_1,t)$. If the just mentioned conditions, in particular the molecular chaos assumption, are satisfied, then

$$\lim_{\varepsilon\to 0} \varepsilon^2 \int_\Delta dx_1 \, \rho_1^\varepsilon(x_1,t) = \int_\Delta dx_1 \, r(x_1,t) \, . \tag{52}$$

So the properly scaled average number of particles can be computed from the solution of the Boltzmann equation.

The actual number of particles in Δ at time t will, of course, differ for different initial configurations of the hard sphere system. Since a probability distribution of the initial configurations is given, the number of particles in Δ at time t is a random variable on Γ. To discuss its properties let me introduce some notation. For a measurable one particle function $f: \Lambda\times R^3 \to R$ let $X^\varepsilon(f)$ be the following sum function on Γ

$$X^\varepsilon(f)\big|_{(\Lambda\times R^3)^n}(x_1,\dots,x_n) = \sum_{j=1}^{n} f(x_j) \tag{53}$$

and let $X^\varepsilon(f,t)$ be the function $X^\varepsilon(f)$ evolved under the hard sphere dynamics with spheres of diameter ε. $X^\varepsilon(f,t)$ is a random variable on Γ,μ^ε, where μ^ε is the initial measure, cf. (33). ($X^\varepsilon(f,t)$ is defined only for μ^ε-almost all points in Γ.) To recall that $X^\varepsilon(f,t)$ depends on ε through the dynamics and through μ^ε, I use the superscript ε. If f is the indicator function of the set Δ ($\chi_\Delta(x_1) = 1$, if $x_1 \in \Delta$, and $\chi_\Delta(x_1) = 0$ otherwise), then $X^\varepsilon(\chi_\Delta,t)$ is the number of particles in Δ at time t.

Let me compute the variance of $\varepsilon^2 X^\varepsilon(f,t)$ for a continuous f of compact support,

$$\mu^\varepsilon(|\varepsilon^2X^\varepsilon(f,t) - \mu^\varepsilon(\varepsilon^2X^\varepsilon(f,t))|^2) = \varepsilon^4\{\int dx_1dx_2\rho_2^\varepsilon(x_1,x_2,t)f(x_1)f(x_2)$$

$$+ \int dx_1\rho_1^\varepsilon(x_1,t)|f(x_1)|^2 - (\int dx_1\rho_1^\varepsilon(x_1,t)f(x_1))^2\}. \tag{54}$$

By (46) and the molecular chaos assumption (47) $\varepsilon^4\rho_2^\varepsilon(x_1,x_2,t) \to r(x_1,t)r(x_2,t)$ as $\varepsilon \to 0$ and therefore the variance of $\varepsilon^2X^\varepsilon(f,t)$ converges to zero as $\varepsilon \to 0$. Thus, the distribution of $\varepsilon^2X^\varepsilon(f,t)$ becomes sharply peaked around $\int dx_1r(x_1,t)f(x_1)$ for small ε. This result offers a neat solution to an old and puzzling problem: In as much as the Boltzmann equation becomes valid, the actual number and the average number of particles in a region Δ at time t coincide. This result offers also an interpretation of the low density limit which is close to Boltzmann's original ideas. For small ε and for a typical initial configuration $(x_1,\dots,x_n)$ of hard spheres of diameter ε (typical with respect to the initial measure μ^ε)

$$X^\varepsilon(f,t)(x_1,\dots,x_n) \simeq \varepsilon^{-2}\int dx_1r(x_1,t)f(x_1) \ . \tag{55}$$

So we have a result about typical initial configurations rather than about average quantities.

On the other hand assume that for an initial measure μ^ε satisfying (C1) and (C2)

$$\lim_{\varepsilon\to 0} \mu^\varepsilon(|\varepsilon^2 X^\varepsilon(f,0) - \mu^\varepsilon(\varepsilon^2 X^\varepsilon(f,0))|^2) = 0 \tag{56}$$

for all continuous f of compact support. Then [1] one shows that necessarily

$$r_n(x_1,\ldots,x_n) = \prod_{j=1}^{n} r_1(x_j) . \tag{57}$$

Thus no fluctuations in the scaled number of particles in arbitrary regions is equivalent to the assumption of molecular chaos. Lanford's theorem implies that if initially the scaled number of particles does not fluctuate, it will also not fluctuate at a later time.

To study the fluctuations in more detail one has to magnify them tremendously: One subtracts out the average value and scales only proportional to ε^{-1}. Therefore I define the <u>fluctuation field</u>

$$\xi^\varepsilon(f,t) = \varepsilon\{X^\varepsilon(f,t) - \mu^\varepsilon(X^\varepsilon(f,t))\} . \tag{58}$$

<u>Conjecture</u>. Under further assumptions on the initial state μ^ε and on f

$$\xi^\varepsilon(f,t) \to \xi(f,t) \tag{59}$$

as $\varepsilon \to 0$, where $\xi(f,t)$ is a Gaussian random field.

This means that the joint distribution of $\xi^\varepsilon(f_1,t_1),\ldots,$ $\xi^\varepsilon(f_n,t_n)$ converges weakly to a Gaussian distribution.

On a formal level this problem has been studied in the physical literature [17 - 20] and rigorously for a underlying stochastic dynamics in [21 - 23]. It is believed that the Gaussian random field $\xi(f,t)$ has the following structure. If one write formally $\xi(f,t) = \int dq dp f(q,p)\xi(q,p,t)$, then $\xi(q,p,t)$ should satisfy the linear, time-dependent stochastic differential equation

$$\frac{\partial}{\partial t}\,\xi(q,p,t) = (L_{r(t)}\xi)(q,p,t) + F(q,p,t)\ . \tag{60}$$

$L_{r(t)}$ is the time-dependent linear operator obtained by linearizing (49) at $r(t)$, i.e.

$$L_{r(t)}\xi(q,p) = -p\cdot\nabla_q\xi(q,p) + \frac{t}{|t|}\int\limits_{\omega\cdot(p-p_1)\geq 0} dp_1\, d\omega\ \omega\cdot(p-p_1)$$
$$\{r(q,p',t)\xi(q,p_1') + r(q,p_1',t)\xi(q,p') - r(q,p,t)\xi(q,p_1) \tag{61}$$
$$- r(q,p_1,t)\xi(q,p)\}\ .$$

$F(q,p,t)$ is a Gaussian white noise fluctuating force with mean zero and covariance

$$\int dp d\overline{p}\ g(p)f(\overline{p})\ \langle F(q,p,t)F(\overline{q},\overline{p},\overline{t})\rangle = \frac{1}{2}\delta(t-\overline{t})\delta(q-\overline{q})$$

$$\int dp \int\limits_{\omega\cdot(p-p_1)\geq 0} dp_1\, d\omega\ \omega\cdot(p-p_1)r(q,p_1',t)r(q,p',t) \tag{62}$$
$$\{g(p_1') + g(p') - g(p_1) - g(p)\}\{f(p_1') + f(p') - f(p_1) - f(p)\}.$$

In one particular instant one can even support the physical picture by a proof. One chooses for μ^ε the grand canonical equilibrium distribution of a system of hard spheres of diameter ε at inverse temperature β and fugacity

$$z_\varepsilon = \varepsilon^{-2}z\ . \tag{63}$$

(This corresponds to increasing the density as ε^{-2}.) In this case the distribution of $\varepsilon^2X^\varepsilon(f,t)$ concentrates at $\int dqdpzh_\beta(p)f(q,p)$. (The system approaches the state of an ideal gas as $\varepsilon \to 0$.) As regards to the fluctuations one can prove [4] that at least the covariance of the fluctuation field converges as $\varepsilon \to C$

Theorem. Let $f,g \in L^2(\Lambda\times R^3, zh_\beta(p)dqdp) \equiv \mathcal{H}$ with scalar product $\langle\cdot|\cdot\rangle_{\beta,z}$ and let $0 \leq t - s \leq t_o(ez,\beta)$. Then

$$\lim_{\varepsilon\to 0} \langle\xi^\varepsilon(f,s)\xi^\varepsilon(g,t)\rangle_{\beta,z_\varepsilon} = \langle g|e^{L(t-s)}f\rangle_{\beta,z} , \tag{64}$$

where L is the linearized Boltzmann operator defined by

$$(Lf)(q,p) = -p\cdot\nabla_q f(q,p) + \int_{\omega\cdot(p-p_1)\geq 0} dp_1\, d\omega\, \omega\cdot(p-p_1)\, zh_\beta(p_1) \{f(q,p_1') + f(q,p') - f(q,p_1) - f(q,p)\}. \tag{65}$$

($\{e^{Lt}|t \geq 0\}$ is a contraction semigroup on $\mathcal{H}$ [24,25].)

If one assumes the limiting fluctuation field to exist and to be Gaussian, then its covariance (64) implies that it is indeed of the form (60) with $r(t) = zh_\beta$.

For the mean field limit of an interacting particle system Braun and Hepp [8] have proved, under mild conditions on the interaction potential and the sequence of initial states, that the fluctuation field converges to a Gaussian random field as $\varepsilon \to 0$. Again the limiting field has the conjectured structure (60). $L_{r(t)}$ is now the linearized Vlasov operator. However, since in the limit no collisions between particles occur, the fluctuating force is identically equal to zero.

Let me summarize: One expects that for small ε

$$\varepsilon^2 X^\varepsilon(f,t) \simeq \int dx_1 f(x_1) r(x_1,t) + \varepsilon\xi(f,t) . \tag{66}$$

The scaled number of particles in a region Δ evolves, to first order, deterministicly according to the Boltzmann equation and its fluctuations around the deterministic path are Gaussian. This picture is quite familiar from other areas of statistical physics, cf. e.g. the survey article by van Kampen [26].

References

[1] O. E. Lanford, Time evolution of large classical systems. In: Dynamical Systems, Theory and Applications, ed. J. Moser, Lecture Notes in Physics 38, p. 1. Springer, Berlin (1975)

[2] O. E. Lanford, On the derivation of the Boltzmann equation. Soc. Math. de France, Astérisque 40, 117 (1976)

[3] F. King, BBGKY hierarchy for positive potentials. Ph. D. Thesis, Dep. of Math., Univ. of Cal. at Berkeley (1975)

[4] H. van Beijeren, O. E. Lanford, J. L. Lebowitz, H. Spohn, Equilibrium time correlation functions in the low density limit, preprint (1979)

[5] H. Spohn, Kinetic equations from Hamiltonian dynamics: The Markovian limit. Lecture Notes, University of Leuven, 1979

[6] G. Papanicalaou, private communication

[7] H. Neunzert, Neuere qualitative und numerische Methoden in der Plasmaphysik, Vorlesungsmanuskript Paderborn, 1975

[8] W. Braun, K. Hepp, Comm. Math. Phys. 56, 101 (1977)

[9] G. Gallavotti, Lectures on the billiard. In: Dynamical Systems, Theory and Application, ed. J. Moser. Lecture Notes in Physics 38, p. 236. Springer, Berlin (1975)

[10] G. Papanicalaou, Bull. AMS 81, 330 (1975)

[11] E. H. Hauge, What can one learn from Lorentz models? In: Transport Phenomena, ed. L. Garrido, Lecture Notes in Physics 31, p. 338. Springer, Berlin (1974)

[12] B. J. Alder, W. E. Alley, J. Stat. Phys. 19, 341 (1978)

[13] J. C. Lewis, J. A. Tjon, Physics Letters 66A, 349 (1978)

[14] R. K. Alexander, The infinite hard sphere system. Ph. D. Thesis, Dep. of Math. Univ. of Cal. at Berkeley (1975)

[15] C. Cercignani, Transport Theory and Statistical Physics 2, 211 (1972)

[16] O. E. Lanford, Notes of the lectures at the troisième cycle at Lausanne, 1978, unpublished

[17] A. A. Abrikosov, I. M. Khalatnikov, Soviet Phys. JETP

34, 135 (1958)

[18] M. Bixon, R. Zwanzig, Phys. Rev. 187, 267 (1969)

[19] R. F. Fox, G. E. Uhlenbeck, Phys. Fluids 13, 1893, 2881 (1970)

[20] T. Kirkpatrick, E. G. D. Cohen, J. R. Dorfman, Phys. Rev. Lett. 42, 862 (1979)

[21] N. G. van Kampen, Phys. Lett. A13, 458 (1976)

[22] M. Kac, J. Logan, Fluctuations. Studies in Statistical Mechanics, to appear

[23] M. Kac, Fluctuations near and far from equilibrium. In: Statistical Mechanics and Statistical Methods in Theory and Application. Ed. U. Landman. Plenum Press, New York (1977)

[24] M. Klaus, Helv. Phys. Acta 48, 99 (1975)

[25] R. Ellis , M. Pinsky. J. de Mathématique Pure et Appliqué 54, 125 (1975)

[26] N. G. van Kampen, The expansion of the master equation. In: Advances in Chemical Physics 34, p. 245, ed. I. Prigogine. John Wiley, New York (1976)

MICROSCOPIC DERIVATION OF THE BOLTZMANN EQUATION

H. Spohn
Universität München, München, F.R.G.

Up to date (1987), I list references on the rigorous microscopic derivation of the Boltzmann equation from the mechanical model of hard sferes with elastic collisions, resp. of classical particles interacting through a smooth pair potential.

The basic references are:

[1] O.E. Lanford, Time evolution of large classical systems, in: Dynamical Systems and Applications, ed. J. Moser, Lecture Notes in Physics Vol. 38, 1 - 111. Springer, Berlin, 1975.

[2] O.E. Lanford, On a derivation of the Boltzmann equation, Astérisque, **40**, 117-137 (1976) reprinted in

[3] Nonequilibrium Phenomena I, The Boltzmann Equation, eds. J.L. Lebowitz and E.W. Montroll, Studies in Statistical Mechanics X, North-Holland, Amsterdam, 1983.

[4] O.E. Lanford, Notes of the lectures at the troisième cycle at Lausanne, 1978, unpublished.

[5] O.E. Lanford, Hard-sphere gas in the Boltzmann-Grad limit, Physica **106A**, 70-76 (1981).

[6] F. King, BBGKY hierarchy for positive potentials, PH. D. thesis, Dept. of Mathematics, Univ. of California at Berkeley, 1975.

A technical step missing in these works is completed in

[7] R. Illner and M. Pulvirenti, A derivation of the BBGKY-hierarchy for hard sphere particle systems, preprint Univ. of Victoria, 1985, to appear in Transport Theory and Statistical Physics.

[8] H. Spohn, On the integrated form of the BBGKY hierarchy for hard spheres, preprint.

[9] K. Uchiyama, On the derivation of the Boltzmann equation from a deterministic motion of many particles, Taniguchi Symp. PMMP Katata 1985, p. 421-441.

[10] K. Uchiyama, Derivation of the Boltzmann equation from particle dynamics, preprint, 1987.

The technical machinery developed in [1] and [6] has been used to tackle related problems.

(i) fluctuation theory for the Boltzmann equation.

[11] H. van Beijeren, O.E. Lanford, J.L. Lebowitz, H. Spohn, Equilibrium time correlation functions in the low density limit, J. Stat. Phys. **22**, 237-257 (1980).

[12] H. Spohn, Fluctuations around the Boltzmann equation, J. Stat. Phys. **26**, 285-305 (1981).

[13] H. Spohn, Fluctuation theory for the Boltzmann equation, in [3].

[14] H. Spohn, Boltzmann equation and Boltzmann hierarchy. In "Kinetic Theories and the Boltzmann Equation", ed. C. Cercignani, Lecture Notes in Math. 1048, p. 207-220, Springer, Berlin, 1984.

(ii) the Lorentz gas

[15] G. Gallavotti, Rigorous theory of the Boltzmann equation in the Lorentz gas, Nota interna n. 358, Univ. di Roma and Phys. Rev. **185**, 308 (1969).

[16] H. Spohn, The Lorentz process converges to a random flight process, Comm. Math. Phys. **60**, 277-290 (1978).

[17] H. Babovsky, Diplomarbeit, Univ. Kaiserslautern, 1980.

[18] J.L. Lebowitz and H. Spohn, Steady state self-diffusion at low density, J. Stat. Phys. **29**, 39-55, (1982).

[19] C. Boldrighini, L.A. Bunimovich, Y.G. Sinai, On the Boltzmann equation for the Lorentz gas, J. Stat. Phys. **32**, 477 (1983).

(iii) expanding gas in two and three space dimensions

[20] R. Illner and M. Pulvirenti, Global validity of the Boltzmann equation for a two-dimensional rare gas in vacuum, Comm. Math. Phys. **105**, 189-203 (1986).

[21] M. Pulvirenti, Global validity of the Boltzmann equation for a three-dimensional rare gas in vacuum, Comm. Math. Phys. **113**, 79-85 (1987).

[22] J.L. Lebowitz and H. Spohn, On the time evolution of macroscopic systems, Comm. Pure Appl. Math. **36**, 593-613 (1983).

(iv) hard spheres with stochastic collisions

[23] C. Cercignani, The Grad limit for a system of soft spheres, Comm. Pure Appl. Math. **36**, 479-494 (1983).

(vi) non-validity of the Boltzmann equation for hard squares

[24] K. Uchiyama, On the Boltzmann-Grad limit for the Broadwell of the Boltzmann equation, preprint, 1987.

As such the results are on a satisfactory level with one outstanding problem remaining unsolved: the validity of the derivation global in time.

There is an extensive physical literature on the derivation of the Boltzmann equation and extensions to moderate densities.

Printed in the United States
By Bookmasters